HIGH SPEED FLIGHT

HIGH SPEED FLIGHT

BILL SWEETMAN

JANE'S

Copyright © Bill Sweetman 1983

First published in the United Kingdom in
1983 by
Jane's Publishing Company Limited
238 City Road, London EC1V 2PU

ISBN 0 7106 0196 4

Distributed in Canada, the Philippines and
the USA and its dependencies by
Science Books International Inc
Russia Wharf, 286 Congress Street
Boston, MA 02210

ISBN 0 86720 661 6

Designed by Peter Matthews

Printed in the United Kingdom by
Netherwood Dalton & Co Ltd
Huddersfield

Contents

For Mary-Pat

0-150mph

Those magnificent men

0–150mph

Those magnificent men

THE aircraft of aviation's formative years were slow, to an almost incredible extent, and people at large were accordingly unaware of the potential of controlled flight. The courage and determination of the pioneers of flying are rendered all the more remarkable by the fact that there was in all honesty no convincing reply to the key question: "Even if it does fly, what is it for?" What, indeed? Their aeroplanes were slower than contemporary railway trains. Worse, the early aircraft were not a great deal faster than quite commonly encountered winds; while a train could forge ahead powerfully against a headwind, an aircraft's speed over the ground would be halved. Its endurance in the air being a constant, its range would also be halved. Small wonder that even a well informed populariser of science such as H. G. Wells, in *War in the Air*, saw the aeroplane as a flimsy and dangerous appendage to the airship: and Wells' book was published in 1908, when a number of workers had already mastered the essentials of heavier-than-air flight. These essentials are rightly summed up as "sustained, powered and controlled flight"; but to be more than a fairground stunt, flying had also to be swift. Otherwise, why trust one's life, or one's military security, to such a machine?

It took little more than a decade for the aeroplane to silence the carping multitudes. It was no coincidence that the world acknowledged the aeroplane to be practicable at about the same time as aircraft were built that could cover more than a hundred miles in an hour, irrespective of tracks, roads or terrain. The advantages of such sheer speed in attack or reconnaissance were too clear for even the most blinkered cavalry officer to ignore. In a more technical sphere, the achievement of such speeds by aeroplanes put the heavier-than-air machine in an altogether different category from the stately but slow airship.

Such public and official acceptance of the aeroplane marked an end to a unique and fascinating phase in the development of the aeroplane, a phase in which progress was at once restricted and encouraged by the meagre resources available to further the advance of technology. Restricted, because an idea could seldom be explored much faster or much farther than one inventor could manage on his own resources; encouraged, because there was no "officially favoured" solution to the problems of flight, and everybody in the game had an equal chance.

The pioneers threw their slight resources against a phalanx of problems which the engineering achievements of their time had largely bypassed.

The disciplines in which they made the first steps have remained in many cases either unique to the aerospace industry or technically dominated by aviation. Advances in each of these main areas have been related to advances in aircraft design as a whole, and in that sense progress in each of them has been co-ordinated. However, this co-ordination has been rough-and-ready at best. When one area of development leaps ahead of the others, there ensues a rapid process of catching up which is equally productive of good and bad ideas. But in the year 1800 there was no problem of this sort, because there was no serious work at all in any of the necessary areas of research.

The most fundamental of the problems facing those who thought of flying machines was the relationship of the machine to the air. Birds demonstrated clearly that something that weighed more than air could be made to fly, but the brief and fatal trajectories of many birdmen had proved that the secret was something more subtle than simply strapping on a frame of wood and feathers and waving one's arms about. It was only in the 1800s that investigators even began to realise just how delicate a balance had to be maintained if a manmade flying object was to achieve a passable imitation of controlled flight. Since the only previous discipline bearing any relationship to the new science of aerodynamics was confined to the empirical techniques involved in the shaping of boat hulls – the only other manmade bodies designed to pass through a fluid – the pioneers of flight were facing a complete void.

They were equally alone in their endeavours to give their ideas physical form. The nineteenth-century engineering mind was not greatly preoccupied with mass. The tonnage of iron that went into the Victorians' creations was a subject for pride, not a necessary evil; iron, coal, and their offspring, steel, the staples of the age, were cheap and tending to become cheaper. The vast bulk of engineering research was based on such abundance, took no interest in structures that were light and efficient, and equated light weight with flimsiness.

In only one aspect of their endeavours did the early fathers of the aeroplane receive any substantial help from outside their fraternity. This was in their search for something which could store and then release sufficient energy to propel their craft. The railway engineers of the 1800s made enormous strides in the technology of propulsion; the trouble was that they concentrated entirely on the steam engine. Fortunately for the aviators, the steam engine was almost as ill-suited to the dream of a self-propelled road vehicle as it was to the aeroplane. Therefore, there were two independent groups of researchers in pursuit of an answer to similar propulsion problems.

Despite the assistance given to aviation by the developers of lighter, less cumbersome engines, the odds have always been against the pioneer who

Wilbur and Orville Wright's painstaking path to flight progressed through kite and glider experiments to powered aircraft of similar design. Their success with the biplane layout was a significant factor in its early dominance of aviation *(Flight International)*

attempts to tackle too many new areas of knowledge simultaneously, and they are especially unfavourable when his resources are limited. Even though the aeroplane designs of William Samuel Henson and John Stringfellow are to the modern eye quite recognisable as aeroplanes, and even though most of the elements of a modern aircraft are there, the fact remained that their ambitious schemes of international air routes using large steam-powered monoplanes were much too far beyond the contemporary state of the art to be remotely feasible. The ridicule which greeted their more adventurous attempts at advertising their project was quite well founded in common sense.

The successful, constructive and best remembered pioneers of the 1800s – such as Sir George Cayley, Lawrence Hargrave and Otto Lilienthal – set aside the problem of power and used gliders as their research tools. Freed from the trouble and cost of powerplants, designers like Lilienthal and Hargrave, Octave Chanute of America and Percy Pilcher in Britain could make repeated flights; they could easily modify and replace their cheap, unpowered vehicles and make a series of experimental flights to determine the effects of the modifications; in short, they could steadily accumulate answers to the questions of stability, control and structure design.

Meanwhile there appeared an engine to meet the aviators' needs. This was the Otto four-stroke petrol engine; its fundamentally important advantage over the steam engine was that its working fluid was air, which could be breathed in from the atmosphere, rather than water, which had to be carried within the vehicle. The first airborne use of the Otto engine, by the airship designer Wölfert in 1897, ended in a fatal crash. Despite this inauspicious start the Otto engine's compactness and light weight had already attracted the attention of many aviators. Pilcher was preparing a powered aeroplane when he was mortally wounded in his Hawk biplane; his work was closely followed by the Wright brothers, who took the same route from gliders to a powered aircraft.

The dominance of glider research had very important effects on the early technical development of the aeroplane. As well as easing the transition to powered flight, glider research encouraged concentration on low airspeeds. In the same way that the modern hang-glider – a slow, cheap and primitive unpowered aircraft – evolved into a new tribe of slow, cheap and primitive "microlite" powered aircraft, the gliders of the Wright brothers evolved into slow powered aeroplanes.

Like a modern microlite, the early powered aeroplanes of the Wrights had engines of relatively low output, but made up for this deficiency by being light in weight and having a great deal of wing area. To combine low weight and large size, the Wrights (influenced to some extent by the railway engineer Chanute) adapted a bridge truss to their needs. One exception to the extravagant supplies of iron enjoyed by the Victorian engineers had been the American railways, where bridges had to be built far from any centres of iron or steel production. A process of innovation, trial and (occasionally) disastrous error had generated a number of more economical structural concepts, including trusses which included some members in tension and were based on the use of wood and cable. This formed the basis for the braced biplane, and in large measure the immense success of the Wright brothers made the biplane popular among other pioneers in Europe and America. Practical but slow biplanes, sauntering along at something less than a mile a minute, became the standard in Europe and the USA.

The first aircraft to lift themselves above the average speed of the day owed their high performance to a remarkable new engine, one of the first internal-combustion engines to depart from automobile practice. Before 1907 there was no such thing as a production-type aero-engine. The French Antoinette was about the best available, generating 50hp from 210lb weight; it was based on an engine which had been originally designed for use in speedboats. It was no great advance on the best aero-engine of 1903: a highly advanced design built by Charles Manly. As an engine, Manly's design was successful. It had the misfortune of being attached to a basically unsound airframe, the tandem-winged Aerodrome designed by Professor Samuel Langley for the Smithsonian Institution. The Aerodrome's two ventures into the air consisted of brief dives from its launching rail into the Potomac River.

Manly's engine, however, contained a number of new features. Its five cylinders were arranged like the spokes of a wheel, or the points of a star, in a pioneer example of the radial layout. Manly's solutions to some of the problems of the radial were prophetic: they included the use of a master con-

As a sport of wealthy — and, in the general view, insane — amateurs, flying was quick to develop a competitive side. The sport of pylon racing still flourishes in the United States, seventy years after meetings like this started in Europe (Smithsonian Institution)

necting rod on one piston, carrying an annular big end to which the other "slave" connecting rods were attached.

The engine fitted to the Aerodrome was water-cooled, as was the Antoinette. Direct air cooling offered a weight saving for a given power output, but was unreliable. In 1907, however, the French engineer Laurent Séguin designed an engine in which the problems of cooling were overcome. Séguin's solution was revolutionary, in more than one sense. It was a radial engine in which the crankshaft was rigidly bolted to the airframe and the body of the engine revolved with the propeller.

The rotary engine, as Séguin's new Gnôme and its many imitators were called, was inherently smooth-running by virtue of its constantly whirling mass. It could be run reliably at high power outputs because of the excellent cooling offered by the rapid movement of the cylinders through the air. What assured it of a place on board the fastest aircraft of the years before the First World War, however, was the fact that an engine of doubled output could be produced simply by joining two sets of cylinders in tandem on a common crankcase and crankshaft.

The rotary was the best engine of its day, even though it always suffered from the problems that were eventually to terminate its career. Accurate control of the amount of fuel and air reaching the cylinders was rendered virtually impossible because of the high centrifugal forces within the rotating engine; the fuel/air mixture was sprayed into the crankcase and allowed to find its way outwards through the induction system to each cylinder. In the absence of precise throttle control and stable manifold conditions, the aircraft designers instead installed an ignition switch on the top of the control column, so that the engine could be rapidly switched on and off if less than full power was required at short notice. This unorthodox procedure was known as "blipping", but was not the worst of the pilot's worries. The spinning engine acted as a huge gyroscope, and like any such instrument strongly resented being forced out of its accustomed plane of rotation. If the pilot forced the aeroplane one way, the engine would promptly attempt to turn at right angles to the aeroplane. If the pilot wanted to retain any measure of control over his attitude and direction, he had to anticipate the kick of the engine and make a simultaneous correction. Some skilled combat pilots learned to make the rotary help them in spite of itself; if hard pressed, they would stimulate a gyroscopic reaction and then follow it with the controls rather than fight the engine, producing a manoeuvre which no conventionally powered aircraft could follow. In general, however, the gyroscopic behaviour of the rotary engine was a dangerous nuisance. To add insult to injury, the rotary could be lubricated only by the one-way passage of castor oil through its system; any more complex closed-circuit lubrication was out of the question because of the forces which threw all fluids within the engine outwards. Some of the castor oil was thrown clear of the aircraft, but far too much of it was absorbed by the pilot, with inevitable and embarrassing results.

But the performance of aircraft with the rotary engine, and particularly with its two-row derivatives, was worth all the trouble. In August 1909 American pioneer Glenn Curtiss had won the first Gordon-Bennett race at Rheims, at just over 47mph; within a little more than two years there were two design teams working on aircraft which would be twice as fast, or even faster, and both used Gnômes.

The first of these to appear, and the most famous, was the Deperdussin racer. Designed by Louis Bechereau, the Deperdussin represented a welcome acknowledgement of the fact that air was a fluid that could create drag as well as lift. Bechereau brought the pilot, the controls and the engine within a single smoothly contoured fuselage. Although some of the Deperdussin racers built in 1912–14 had fuselages constructed on an internal frame, the type is rightly associated with the "monocoque" (literally, single-shell) method of construction. The monocoque bodies were built slowly by hand, being formed around a solid mould from thin strips of tulip wood. The strips of wood were glued in place layer by layer, and were covered with a protective coat of fabric inside and out, to create a light, strong and streamlined shell only a fraction of an inch thick.

The Deperdussin's wings were also more streamlined in form than the cumbersome structures of its contemporaries. To begin with, it was a monoplane. Despite the dominance of the biplane, designers such as Louis Blériot and Levasseur had achieved great success with monoplanes, and Bechereau's design built on their work. The Deperdussin wing had no direct bracing struts. It was braced against negative and positive "g" loads by wires above and below, the upper (landing wires) set being run from a compact assembly of streamlined struts set above the forward fuselage.

The later Deperdussin racers were powered by the 160hp twin-row, 14-cylinder Gnôme. Bechereau installed the engine in a cowling which not only helped to streamline the aircraft but also caught some of the flying castor oil, and as a last refinement the designer added an elegant conical fairing to the propeller boss – the first "spinner" to be used on a radial engine. With a combination of high power and aerodynamic excellence, the Deperdussins were unbeatable on the racecourse and in the record books right up to the outbreak of war – provided that they stayed in the air. This was usually the case in land-based races, but the fragile and highly strung racers were less well suited to the rougher conditions of seaplane racing.

In September 1913 Maurice Prevost's Deperdussin won the last pre-war Gordon Bennett race at an average of 124·5mph. At that time there was only one aircraft under design that was potentially as fast

11

Louis Bechereau's elegant Deperdussin marked a great leap forward in aerodynamic refinement, with its spinnered propeller and snugly cowled Gnôme rotary engine faired into a smooth, slender body handcrafted from tulip wood. Note also the discs over the wheels, replacing plain spokes *(Smithsonian Institution)*

The only way to lubricate the Gnôme was by a one-way system using castor oil. Here a ground crew get a laxative showerbath as they hold back a Deperdussin on an engine run. The pilot got similar treatment throughout the flight *(Smithsonian Institution)*

or faster; it was a closely guarded secret at that time, before its first flight, and not a great deal is known about it today.

Its design origins went back to the summer of 1912, when two young designers, H. P. Folland and Geoffrey de Havilland, at the Royal Aircraft Factory at Farnborough, England, started work on a light military aircraft using the powerful two-row Gnôme. Originally designated B.S.1, the initials standing for Blériot Scout (the War Office used the French designer-pilot's name to indicate any aircraft with its engine and propeller at the front), the aircraft was known as the S.E.2 (Scout Experimental 2) when completed and flown. It was a biplane with a semi-monocoque fuselage, in which a rigid outer skin was supported by internal stiffening members. The S.E.2 managed a respectable 92mph from its 100hp Gnôme before it was wrecked in March 1913, and set Folland working on the design of a more advanced aircraft. The S.E.2's successor would be aerodynamically refined to take advantage of the 160hp Gnôme: by February 1914, as the S.E.4 took shape, it was expected to be the fastest aircraft in the world.

The S.E.4 was extremely clean despite its biplane configuration. Folland reduced the bracing struts to the bare essentials, with a single interplane I-strut on each side. The ends of the struts were flared out to pick up the front and rear spars. Bracing wires were kept to a minimum by the simple layout, and the S.E.4 was among the first aircraft to use the new streamline-section wire developed by the Royal Aircraft Factory. ("Rafwire" was developed despite the belief of many practical-minded aviators that the shape of so small an object as a bracing wire could not possibly be important. It was, and Rafwire and similar products became universally used.) The design of the lifting surfaces themselves was astonishingly advanced: the S.E.4 was fitted with full-span flap/ailerons on the upper and lower wings. All four surfaces could be drooped together to improve the flying characteristics at low speeds, and in high-speed flight the flaps could even be deflected slightly above their normal position to reduce drag – a system which is used on the Learfan 2100, Lockheed U-2 and a few other advanced high-efficiency aircraft of the present day.

In general streamlining the S.E.4 echoed the Deperdussin, with a rigid-skinned fuselage and a closely cowled and spinnered engine installation. Folland introduced an added feature – a clear celluloid cover over the cockpit, a device never seen before. It was too novel for the RFC test pilots, who had the cover removed before flying the S.E.4. Before the S.E.4 was wrecked, in the month that war broke out, it had attained 135mph, slightly more than the official world's air speed record.

Not only were the Deperdussin and the S.E.4 aerodynamically cleaner than any of the warplanes produced in the conflict that ended the development of specialised high-speed aircraft, but they flew faster than any of the operational types produced before the final year of the war. Far from stimulating technical development, the main effect of military demands was to add weight in the form of guns and other weapons. Aerodynamic refinement was almost entirely subordinated to ease of manufacture. In the world of the scout pilots, straight-line speed was less valued than the qualities summed up in the adjective "splitarse": applied to any fighter aircraft, the word implied high manoeuvrability at low airspeeds, together with tolerance of the most powerful control inputs that a pilot in mortal danger could muster.

Despite the low priority given to speed, and the accumulation of weight aboard the average fighter, there was some increase in aircraft speed during the four years of war. Much of this was due to a

In the S.E.4 biplane Folland and de Havilland of Farnborough nearly equalled the grace of the Deperdussin. Features such as its single wing struts were not copied in production until the 1930s, while its remarkable system of flap/ailerons was years ahead of its time

renewed, military-inspired round of engine development. One of the first moves in this direction had started in January 1912. In commemoration of the Kaiser's birthday, the German war office offered a prize for the best German-designed aero-engine. The Daimler motor company was a favoured entrant, having already (in 1911) started production of an in-line six-cylinder aero-engine using the same combination of separately formed steel cylinders and welded-sheet-steel water jackets as their car engines. Like the cars produced by the company, these aero-engines bore the name Mercedes.

Not only were the Mercedes-Benz aero-engines widely adopted by the German armed forces, but they were also highly influential abroad. Under pressure of war Rolls-Royce went into the manufacture of aero-engines in 1914 (the decision to do so was not an easy one, since the firm's co-founder, C. S. Rolls, had been killed in an air crash in 1911). Although the first Rolls-Royce aero-engine, the six-cylinder Hawk, was based on the powerplant of the famous Silver Ghost car, it embodied a number of features introduced on Mercedes racing and aircraft engines. Although the Hawk was mainly confined to use on small airships and training aircraft, it provided the upperworks for the V-12 Eagle, one of the most successful high-powered engines of its day. From that time onwards, variations on the V-12 theme became increasingly common in the aircraft industry.

These engines, and other designs such as the V-8 Hispano-Suiza engine, gradually ended the rotary's brief reign as a powerful, irascible and occasionally treacherous tyrant of the skies. By the end of the war only the newly formed Royal Air Force continued to make extensive use of rotaries, Bentley's ultimate variations on the theme being fitted in the Sopwith Snipe.

In one vitally important respect, however, the engine manufacturer's job became far more difficult towards the end of the war. The problem was the fuel with which the air forces were being supplied. In 1914 most people considered that petrol was petrol, and that was all: the vital differences between various kinds of hydrocarbons were only dimly perceived. In particular it was not realised that gasoline from different parts of the world varied widely in its tendency to explode in the cylinders of an engine before the pistons reached the top of the compression stroke, and many engines were being lost due to this destructive "knocking". The problem was closely linked to the design of cooling systems, because uneven distribution of heat within the cylinders and locally high temperatures encouraged detonation. Some widely produced engines had to be "fuel-cooled" in service: that is to say, they had to be run on rich fuel mixtures to prevent them from detonating, but this wasted fuel and reduced output.

The scientific investigation into the contribution of fuel chemistry to the detonation problem was just beginning when the United States entered the war.

An indirect and quite unforeseeable result of the new alliance was that the engines of many Allied aircraft began to run rough, and in some cases broke their crankshafts and failed completely. The reason pointed the way to an important line of developments in aircraft propulsion. Before the entry of the US into the war the British and French had been using gasolines from wells in Borneo, Java and Sumatra. It was all "straight-run" fuel, distilled directly from the products of a single well, and was occasionally spiked with benzol (derived from coal tar) when this was found to improve engine running. There was, at that time, no systematic method of evaluating the resistance of a fuel to knocking. In the USA, the oil industry had tended to rate gasoline in proportion to its specific gravity: fuel from Pennsylvania wells was found to be more dense than that from California, and was accordingly used as the basis for aviation fuel. Sadly, its knock resistance was far lower than that of Californian gasoline, which was among the best to be found anywhere; in 1917, when the USA took on the burden of supplying fuel to its allies, the quality of fuel sharply declined.

Further retrograde steps were made in the application of aerodynamics to design. Of wartime combat aircraft, few equalled the refinement of the Deperdussin or S.E.4. One of the few exceptions was the German Roland C.II two-seat fighter and reconnaissance aircraft. Designed by Dipl Ing Tantzen, the Roland featured an elegantly contoured fuselage built, in the Deperdussin manner, from thin strips of wood laid up and glued on a mould. The aircraft took its popular name of Walfisch (whale) from the shape and depth of its fuselage, which filled the entire gap between the wings. Bracing was confined to a single I-shaped interplane strut on either side, and a small roll-over cage to protect the crew's necks in a nose-over.

However, the main wartime contribution to the aerodynamic art was distinctly negative, in that it greatly reinforced the prejudice against the monoplane configuration for high-performance aircraft. Even such a fighter as the Bristol M.1C was relegated to service with squadrons in the Middle East, despite the fact that it passed its service acceptance tests and attained a highly respectable 132mph top speed. The monoplane was generally considered unsafe, lacking in climb and turning performance, or both. For the modest speeds attainable by warplanes of the First World War there was probably little to choose between the biplane and the monoplane; the problem in later developments was that the speed limitations of the biplane became accepted as a fact of life, and fighter design dropped into a 15-year rut.

At this point, too, the aviator Glenn Curtiss had achieved a speed under controlled conditions which was greater than the existing air speed record – on a motorcycle. Clearly, the aeroplane had some way to go.

150-300mph

Race for the Trophy

150–300mph

Race for the Trophy

BETWEEN the First World War and the Great Depression there occurred an outbreak of international machismo on a scale which the aviation industry never experienced again. Air racing, the sport of rich amateurs in the pre-war years, became a substitute for war; and, as in war, many a participant made the final sacrifice for his country.

It was just as well that government-supported racing became widely accepted, because the development of new technology for use in practical high-speed aircraft lost all impetus with the end of the First World War. Contracts were cancelled and names disappeared from the industry. The military, satisfied that the "war to end wars" had been won, saw no reason to develop new aircraft.

The design of the airframe, it was widely considered, had reached the ultimate point of refinement. In the particular case of fighter aircraft, the mistrust of the monoplane became, if anything, more deeply rooted in the immediate post-war years. A number of bad experiences with monoplanes were at least partly to blame, and most of these can be attributed to the novel structural problems presented by a single aircraft wing. The chief of these was torsion, or twisting; an aircraft wing often twists under load, rather than simply bending. A biplane's wings form a large and lightly stressed torsion box, but the spars of early monoplane wings tended to behave as two parallel independent beams, with very poor resistance to torsion. If a wing started to twist upwards under load (such as a pull-out from a dive) its angle of attack would increase, and so would its lift and the twisting stress. The results were often catastrophic. The wartime Fokker D.VIII suffered a series of accidents. The US Army Air Service tested two monoplane fighters at its engineering base at Dayton in 1922: in March the USAAS's Fokker monoplane folded its wings and crashed during a mock combat with the US-designed Loening PW-2; in October one of the Loening monoplanes did the same. In these circumstances, it is possible to see the prejudice against the monoplane as something more than just fanatical conservatism. Slow, lightly stressed transports were the only fully successful monoplanes of the 1920s.

Aerodynamic refinement of service types rarely reached the levels set by Bechereau, Folland and Tantzen years before. Neither the static radial engine, now gaining prominence, nor the in-line engine was as easy to keep cool as the now superseded rotary, and open flat-plate radiators were the order of the era. The replacements for late-wartime types such as the Snipe and the Spad were barely distinguishable from their predecessors. What performance advantages they had stemmed from a new generation of equipment in front of the firewall, rather than any innovations behind it.

In the second half of the war design work had started on a number of advanced engines, although none appeared in time to be used operationally. Three of the new powerplants went into production in Britain. Napier's Lion was a 12-cylinder engine with three banks of cylinders forming a broad arrowhead shape in front view. By comparison with contemporary service engines, it had a high compression ratio (5·8:1) and ran fast (2,000 rev/min). It was fitted with a reduction gear to cut down propeller speed to manageable levels.

Two other excellent British engines emerged from the shadow of a near-disaster in wartime procurement. British engine production had dropped badly behind demand in 1917, and the Ministry of Munitions was anxious to improve the situation. When a newly formed company, ABC Motors, offered them a static radial engine which would not only combine low weight with high output but was also designed specifically for rapid production, they were overjoyed: so much so that the Ministry of Munitions ordered that all production of other Royal Flying Corps fighter engines should stop before the end of 1918. The Armistice was signed before this policy could be fully implemented. This was fortunate, because the ABC Dragonfly proved to be useless: overweight, underpowered and riddled with severe and unacceptable vibration.

It was just as well that the British Admiralty had sponsored its own engine developments, and that these engines were to prove even more successful than the Lion. One of them was designed by Roy Fedden, chief engineer of the Brazil-Straker company. Brazil-Straker was prevented from designing liquid-cooled engines by the terms of its licensing agreement with Rolls-Royce, whose Hawk and Falcon engines it was producing for the RFC and Royal Naval Air Service. The 14-cylinder, two-row engine designed for the Admiralty was named Mercury, the first of a long line of engines designed by Fedden and his successors to take their names from mythology. In 1918 the company was renamed Cosmos Engines, and in the interests of easing production replaced the Mercury with a nine-cylinder, single-row powerplant, with a potential output of a remarkable 500hp: the Jupiter.

The other Admiralty engine had a similarly tortuous lineage. Its design had started at the Royal Aircraft Factory at Farnborough, as the Raf 8, in 1916. Among its many advanced features was an integral, gear-driven supercharger, designed by James E. Ellor. (Ellor was to remain the world's leading expert on the lightweight, high-power centrifugal blower for the next two decades.) Before the Raf 8 could prove itself, the Royal Aircraft Factory was caught in a political storm; the poor combat record of some of the early British aircraft caused an outcry, and Farnborough turned out to be the

Anthony Fokker's D.VIII was the first production fighter with a "cantilever" monoplane wing free from bracing struts or wires. It was also the last for nearly two decades, as the answers to the structural questions which it posed continued to elude designers. After the D.VIII monoplane wings were for many years confined to slow, lightly stressed transports *(Howard Levy)*

scapegoat. It became the Royal Aircraft Establishment, permanently stripped of its right to build aircraft and engines. The Raf 8 designers moved to private engine-builder Siddeley Deasy of Coventry. The name of the firm was changed to Armstrong Siddeley soon afterwards, and the engine which had been the Raf 8 emerged as the Jaguar. In its initial production form it had an impeller fitted to the crankshaft to help the fuel and air to mix. This was not however a supercharger as originally planned for the Raf 8, because it did not compress the incoming air. From the era of the Jupiter, Jaguar and Lion onwards, the British aircraft industry was never without a choice of water-cooled and air-cooled engines. Usually, engine manufacturers were deadly rivals for commercial and military applications and, often, for racing and record-breaking as well.

The immediate post-war years saw only one advanced aero-engine developed successfully in the United States, but it was a design of tremendous importance. Glenn Curtiss and his company had produced the only really successful aircraft of US domestic design during the war. One of the engine designers with whom he worked most closely was Charles Kirkham, who in 1916 designed a new V-12

aero-engine, the K-12, with an output of 340hp. It won no production contracts, because the US Army Air Service was totally committed to the clumsy, heavy Liberty engine, which had been designed for rapid production by a consortium of car manufacturers and looked it. Every so often the USAAS would pull one of Marc Birkigt's elegant Hispano engines out of a European fighter and substitute the Liberty, invariably with horrible results.

Curtiss continued to make experimental use of the K-12. The engine resembled the Hispano, and differed from most earlier liquid-cooled aero-engines, in that it was of "monobloc" design: each row of cylinders was cast and machined as a single block of metal rather than being built from separate cylinders and water jackets, resulting in a less bulky powerplant using a smaller volume of water to cool it. The angle between the banks of cylinders was also reduced, from the normal 90° to just 60°. Narrow-V configuration and monobloc construction gave the K-12 an impressive ratio of power to frontal area. In early 1918 Curtiss fitted the K-12 to a pair of two-seat fighter prototypes: a triplane, designated Model 18-T Wasp, and the Model 18-B Hornet biplane. Interestingly enough, it was the triplane that proved faster, setting a new world's air speed record of 163mph in August 1918.

It was engines such as these that dominated the gladiatorial air races of the 1920s, at least until the last years of the decade. The first series to begin after the Armistice was at once the greatest of all air races and entirely unique among such contests, because it was confined to marine aircraft. Jacques Schneider was a motor-race backer and balloonist who believed that the future lay with the seaplane. In December 1912 he offered a trophy for an annual seaplane race. Its full name was La Coupe d'Aviation Maritime Jacques Schneider, but in the English-speaking world it became known as the Schneider Trophy.

Schneider intended the Trophy to be a stimulus to the development of practical marine aircraft (Curtiss had made the first take-offs and landings by a truly practical floatplane in January 1911) and the original rules included seaworthiness and water-handling trials. These, in the early races, were sufficiently demanding to eliminate more than one contender by total immersion, but were progressively watered down (to coin a phrase) in the interests of sheer speed and spectacle. One of the rules that remained in force, however, was that a nation that won three races in five years would retain the Trophy for all time.

The pre-war Schneider racers were by no means the fastest aircraft of their day, which was hardly surprising in view of the drag of their floats or hulls.

Post-war contenders, however, began to close the gap, largely due to the almost unlimited runway length offered by a water take-off. This was a gradual process, and it was a few years before the actual lap speeds began to approach the flat-out speed capability of the aircraft.

The Sopwith Schneider, built for the 1919 race, was a good example of the early post-war aircraft. Claimed to be capable of 170mph – slightly faster than the landplane Curtiss 18-T fighter of the previous year – the Schneider was closely similar in size to the Sopwith aircraft of the same name that had won the 1914 race. Its wings had the same chord and interplane gap, and were 3ft shorter in span, but the 1919 racer had a Cosmos Jupiter engine delivering four-and-a-half times as much power as the Gnôme rotary of its forebear. It never showed its paces; the 1919 race was voided after the organisation failed in thick fog. In the two succeeding years Italian aircraft were able to "walk" around the course after the other competitors failed to qualify for the race. A victory in either 1922 or 1923 would have been an outright win for Italy; in 1922 this was averted by British brute force. Reginald Mitchell of Supermarine adapted his little Sea King II fighter amphibian to take a Napier Lion, and the big, reliable engine hauled the less-than-beautiful airframe, renamed the Sea Lion, to a British victory. The 1923 Schneider race, however, was to be very different.

The new factor in 1923 was competition from the

Brute force meets aerodynamics: the inelegant Supermarine Sea Lion is hardly one's idea of a racing aircraft, or of a Reginald Mitchell design, but its great advantage was that it could absorb and exploit the power of the new Napier Lion

United States, and competition of a very high order at that. The US racing scene was dominated by Curtiss, with some competitive spurring from the Wright company, and all the Curtiss racers used V-12 monobloc engines derived from the Kirkham design. The first post-war Curtiss to be designed expressly for competition had been a monoplane using a 435hp C-12 engine and some S-3 Scout components. Two examples were ordered by Texas millionaire S. E. J. Cox to compete in the 1920 Gordon-Bennett race, but one crashed and the other was withdrawn for extensive modification.

Meanwhile, a new series of air races had been sponsored in the USA, in memory of the publisher Joseph W. Pulitzer. The first race was held on Thanksgiving Day, in November 1920, at Mitchel Field, Long Island. The Curtiss-Cox racer, modified to a triplane, came in second to the Air Service contender. The USAAS, however, was not to have a monopoly of the Pulitzer. Well ahead of the following year's race, the US Navy ordered from Curtiss two "prototype fighters", designated CF-1 and CF-2. The basically identical aircraft lacked any provision for armament or other operational equipment, and before they flew the pretence of military purpose was abandoned. Pure racing was their function, and they were redesignated CR-1 and CR-2.

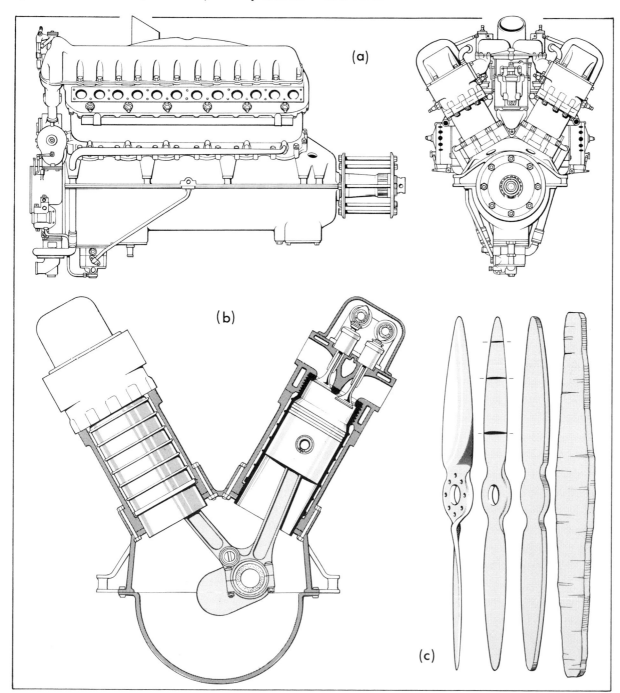

Revolutionary in its day, the Curtiss D-12 engine dominated racing for years and was the inspiration for the later Rolls-Royce V-12s. Slim and compact, with its accessories mounted between the cylinder banks and a 60° V-12 layout (a), the D-12 introduced new internal features such as twin camshafts, and monobloc construction with screwed-in cylinder liners (b). Along with it came Dr S. A. Reed's radical metal propeller, forged, cut, profiled and twisted (c, right to left) from a solid aluminium billet.

The new Curtiss racers were small, clean biplanes of conventional design. Their construction was relatively advanced, with steel-tube fuselage frames and plywood-skinned wings. Where they broke new ground, however, was in their engine installations. The Curtiss CR was fitted with the new Curtiss D-12 engine, housed in a sleek cowling tailored to the V-shape of the cylinder banks. The propeller was based on the designs of Dr S. A. Reed; its blades were thin in section, forged, cut and twisted from solid aluminium alloy, in stark contrast to the elegant carved wooden propellers used on previous aircraft. The Reed propeller could absorb the output of the D-12 better than a larger-diameter, slower-turning propeller, and without the complication of a reduction gear.

The CR-2 won the Pulitzer race in 1921, and went on to set an absolute speed record at 197·8mph a few days later. The Army, however, was not ready to give in to the Navy and followed its rival to Curtiss' door. The Navy had now modified the CR-2 with a new cooling system. In place of the podded Lamblin radiator slung externally between the undercarriage legs, the CR-2 was fitted with a double skin of brass on its wings. These "surface radiators" could, in theory, cool the engine without causing any drag. They were to become almost *de rigueur* for racers and

record-breakers, but were no use at all for any practical application: they were too large, and too difficult to maintain, for commercial use, and endowed a military aircraft with an enormous area in which the smallest bullet-hole could prove rapidly lethal.

The Army's Curtiss R-6 not only had surface radiators but was substantially smaller than the CR-2; spanning just two inches short of 20ft, it weighed 2,120lb for take-off compared with 2,720lb for the earlier aircraft, and had just a plain I-strut between its closely spaced wings. The two little R-6s raised Pulitzer lap speeds above 200mph for the first time, winning the race for the Army Air Service at an average of 205·8mph. Shortly after the race one of the R-6s took the world's air speed record to 224·05mph, and in March 1923 the R-6 raised the mark again, to 236·587mph.

The Navy planned revenge for the 1923 Pulitzer, ordering a pair of R2C-1 racers from Curtiss. Similar in size and weight to the Army's R-6, the R2C-1 was distinguished by the even smaller gap between the wings: the upper set now rested on the fuselage, reducing drag at the price of a poorer view for the pilot and a loss of lifting capability due to the closeness of the wings; the low-pressure zone above the lower wing tended to break down the high-pressure

Graceful but impractical, the last of the American service racers was the Curtiss R3C-2. Features such as an upper wing mounted directly on the fuselage, and the wing-surface radiators, were fine for a racer but were technical dead ends. By 1925, too, even the once radical Curtiss engine was approaching obsolesence *(Smithsonian Institution)*

zone below the upper wing, reducing lift, particularly at low speeds.

By this time the Navy's old CR-1 and CR-2 were outclassed in the Pulitzer contest. As recounted earlier in this chapter, however, speeds in the Schneider seaplane races had not been rising as rapidly as those in the US landplane races; so the US Navy added floats to the Curtiss racers, which became CR-3s. Their team-mate was a reworked version of the Navy-Wright racer which had unsuccessfully challenged the Curtiss grasp on the Pulitzer in the previous year. The NW-2 was heavier than the CR-3s but had the advantage of a larger and more powerful Wright T-3 Tornado engine: the merger between Curtiss and Wright engine interests was some way in the future.

The 1923 Schneider race, held at Cowes because Britain had won in 1922, could have provided the Italians with another attempt at an outright win; despite their defeat of the previous year, a win in 1923 could have given them the necessary three victories in five years. However, it was only the French and British who even attempted to challenge the Americans. It was the best that the British could do with no official support: a revised version of the Sea Lion from 1922, the Sopwith Rainbow – the same aircraft as the 1919 Sopwith Schneider – and the only new contender, the Blackburn Pellet. The last-named was considered the home team's best hope, if only because it was an unknown quantity; with a 450hp Lion, surface-type radiators and a 2,800lb gross weight, it was roughly in the same class as the US Navy racers, and was expected to exceed 160mph. However, its surface radiators proved inadequate and an additional speed-reducing Lamblin radiator was found to be necessary. All that could be said was that the Pellet was probably saved from defeat by sinking during its navigability trials.

Of the non-American entries, only the Sea Lion even completed the course, 20mph slower than the leading CR-3. The Wright challenger crashed at more than 200mph during the race. The winner's average speed was 177·38mph; this was still slower than the Pulitzer landplanes, but was a pointer to the fact that the gap was narrowing. Meanwhile, ever faster times were being recorded by the new Navy racers practising for the Pulitzer Trophy contest; Wright's new TX (later redesignated F2W-1) and the new Curtiss R2C took turns in setting record lap times. The winner was an R2C-1 at 243·68mph; once again, the Wright contender was unlucky and crashed at the end of the race.

The events in the racing scene during 1923, and in particular the outstanding performance by the US Navy Curtiss racers at Cowes, were of great importance. The world learned, in the most dramatic and humiliating fashion, about some of the more advanced work that the United States had done over the previous five years. There were two categories of response – competitive and imitative –

and the effects in both areas were to be felt a decade later.

The direct competitive response did not have the immediate effect that was desired. Britain sponsored an official entry for the 1924 Schneider race, which was to be held at Baltimore. (Possibly because there was as much domestic propaganda value in one US race as in another, the US Navy did not take part in the Pulitzer race that year.) However, H. P. Folland's Lion-powered Gloster II biplane crashed in flight tests, eliminating the only opposition to the US Navy. The service resisted the temptation to claim a flyover victory and put itself within a race of possessing the Trophy for good.

The delay proved fatal for the USN's choice for an outright win. The R2C racer was to be the last Curtiss design specifically intended for the racing circuit, and time was running out for the biplane in the absolute speed stakes. By 1925 the Curtiss biplane was almost universally used by Army and Navy teams, although modified with the 610hp V-1400 development of the D-12. (Standard designations using a letter or letters to denote engine configuration, followed by engine capacity in cubic inches, were to be used for US Navy, Army and US commercial engines.) It was redesignated R3C-1 with wheels and R3C-2 on floats. The floatplane set a world's air speed record for marine aircraft, but the landplane could manage only a 5mph improvement over the previous year's Pulitzer speed.

The R3C-2 won the Schneider race, but only after the most promising of its opponents had been eliminated. This was the Supermarine S.4, developed by Reginald Mitchell. Revealed in August 1925, the S.4 looked like something from another planet alongside the Gloster III, its Folland-designed team-mate. Its lines were unbelievably clean. The monoplane wing was free from all bracing struts and wires, and set at the mid position on a fuselage which streamed smoothly aft from a neatly cowled 700hp Lion. (Napier had steadily modified the Lion, removing accessory equipment between the cylinder banks, to allow tighter cowlings.) One of the first cantilever wings to be used on a high-performance aircraft, the S.4 wing was built in one piece, was skinned in plywood for smoothness and carried full-span flaps and ailerons which could be drooped to reduce landing speed. After a brief career during which it broke the world's speed record for seaplanes (later beaten by the R3C-2) the S.4 crashed during practice at Baltimore. The cause of the accident was not discovered. A high-speed stall caused by an excessively sharp control movement was blamed in some quarters, while flutter was cast as the villain by other observers. Either way, the loss of the S.4 removed the Americans' only serious competition.

The US racers needed only to win in 1926 to end the contest, but following the 1925 Pulitzer race the US Government put a stop to an increasingly expensive and dangerous display of inter-service rivalry

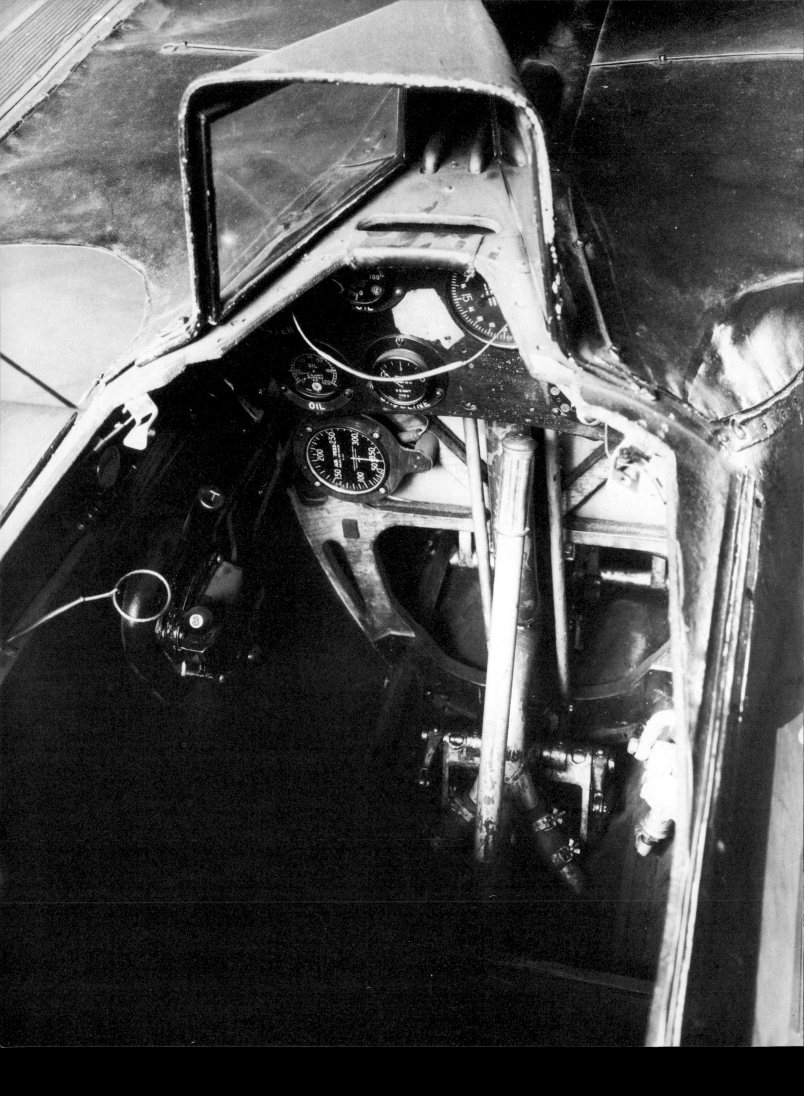

and decreed that no more specialised racers should be built with public funds. The Navy did its best, fitting a bored-out D-12 (Curtiss V-1550) in one R3C, and a Packard engine in another of the older racers. They had one determined and well prepared adversary: Italy. Mussolini had personally ordered that Italy was to win the Trophy, and the engineer Mario Castoldi of Macchi had responded with an aircraft that set the pattern for subsequent racers. The M.39 resembled the S.4 with the exception of the wing, which some held to have been the Achilles heel of the British racer. Castoldi instead designed a low-set, thin-section wing, still made of wood but braced to the floats and the fuselage with steel wire. Fiat bored out its A22 engine to create the 850hp AS2 of similar bulk and weight. All the US Navy pilots could do was hope that the M.39 would be eliminated at some stage in the contest, but it was not to be so; over the Chesapeake Bay course, the Italian team and the M.39 made the Trophy common property again.

The US Navy had lost two aircraft and their pilots in practice for the 1926 race; no doubt this was one of the reasons why the service declined to sponsor an entry for the 1927 contest, run in the sheltered waters off Venice. There was one private entry from the USA, bearing the names of the D-12's designer, Charles Kirkham, and the Navy's star race pilot, Al Williams. A conventional biplane, its most unusual feature was its powerplant – one of the first 24-cylinder aero-engines, the Packard X-2775. Like several manufacturers afterwards, Packard produced the X engine by using modified V-12 compo-

nents on a new crankcase and crankshaft. It developed 1,250hp, but at 45·5lit capacity it was substantially bigger and heavier than its rivals; the result was that the Kirkham-Williams racer was a bulky biplane, in contrast to the sleek new monoplanes.

Italy's main hope for 1927 was the M.52, aiming at a 300mph maximum speed. The aircraft had a slightly smaller wing than the M.39, but was otherwise similar to it; the engine was the AS3, similar to the AS2 but faster running, with greater capacity and a higher compression ratio. In two years the Fiat engineers had developed the 590hp A22 into the 1,000hp AS3.

The Italian team was going to need all that power, because the British were making a determined comeback after their absence in 1926. The Air Ministry, the Royal Aircraft Establishment and the National Physical Laboratory had joined forces with the RAF, two engine manufacturers and three airframe suppliers in a determined effort to bring the Trophy to Britain. Three very different aircraft were developed for the race; this was more than enough, as the rules permitted only three entries per country.

A definite odd-man-out among the British designs was the Short-Bristow Crusader, rapidly christened "Curious Ada" by the pilots of the newly formed High Speed Flight. It was distinguished by its powerplant – an air-cooled radial engine, specially developed from the Jupiter by Roy Fedden's team. It was the first of many radials to challenge the more visually appealing in-line engine on out-

23

right speed. This was just one of many interesting new features in the Mercury, which was as much an exercise in radial design as it was a pure racing engine. The basic design was that of a Jupiter, reduced in capacity by shortening the stroke of the cylinders and pistons. This reduced the frontal area and hence the drag of the engine, at the same time allowing the engine to run faster while keeping the same piston speed.

Even such a high-speed radial would normally deliver too little power in proportion to its frontal area for any chance of success in a sea-level race, so Fedden designed the Mercury to use supercharging at sea level. The supercharger is an air compressor, and had originally been designed to help engines to work at high altitudes; the RAF had taken delivery of its first Siskin IIIA fighters with supercharged Jaguars in 1926. Fedden, however, was the first to use a supercharger as a straightforward means to increase sea-level power. The Mercury's supercharger and its geared drive were more advanced than those of the Armstrong Siddeley Jaguar; in particular, the aerodynamics of the supercharger were based on research by James Ellor of the RAE, who had been working on centrifugal blowers since the days of the Raf 8 design in 1916.

The Mercury was also the first British radial to feature a solid aluminium cylinder head. One of the limitations to the performance of the radial had been the difficulty of designing a cylinder head which would achieve adequate and even cooling despite the need to find room for the valve gear. Earlier Jupiter engines had "poultice" heads, in which an outer aluminium shell was attached to the steel cylinder lining and drew out the heat, as a poultice in medical practice draws out the inflammation from a burn or sore. The solid head did a better job, and by providing more even cooling reduced the risk of detonation, so that the engine could be run safely at higher compression ratios.

Bristol got 960hp out of the Mercury, a very respectable figure for a radial engine with little more than half the capacity of the 1,250hp Packard X-2775. All Fedden's ingenuity, however, was of no avail against the fitter who inadvertently crossed the Crusader's aileron controls when it was reassembled before the race at Venice. The pilot escaped from the first attempt at a practice flight with severe bruises and concussion; history does not record whether the fitter responsible got off as lightly. Experience with the temperamental Mercury was encouraging enough for Bristol to include the supercharger and the cylinder head design in a new generation of Jupiter developments.

Apart from "Curious Ada", all of the British entries were powered by the Napier Lion, designed more than ten years earlier. However, A. J. Rowledge's original design had shown an immense potential for development, and now yielded 900hp in its direct-drive VIIA model, and 875hp with gearing as the VIIB. It was unsupercharged, but

ran at a compression ratio of 10:1 and at a speed of 3,000rev/min. (Comparable figures for the original 450hp engine were 5·8:1 and 2,000rev/min.) Part of this great reserve of power in the original design can be attributed to the basic broad-arrow layout, which yielded a short, rigid engine, and part of it to basic, painstaking engineering skill.

The British fielded two Lion-powered designs, with a mix of geared and direct-drive engines. The Gloster IV was the last biplane to enter the race, representing a further move in the direction of structural cleanliness. The team's main hope was a new design from Reginald Mitchell – the Supermarine S.5, bearing a family resemblance to the S.4 but almost completely different in detail. Mitchell followed Castoldi in a switch to a wire-braced low wing, of wooden construction with metal surface radiators of the low-drag American type. The long, slender fuselage was an aluminium alloy structure of "semi-monocoque" type, with a skin stiffened by internal ribs and stringers. Oil-cooler radiators covered the sides of the fuselage. In an unconventional but logical touch, the starboard float was installed slightly further from the fuselage than the port float, in an effort to counter the powerful torque from the engine. The S.5 proved substantially faster than the Gloster IV, and the British team for the race itself comprised two S.5s and a single Gloster biplane.

Their opposition was the Italian team of two M.52s and one of the older M.39s. The Kirkham-Williams racer could not meet the date set for qualifying entries, and the British objected to any extension of the deadline; the lure of the Trophy was too powerful for the British sense of sportsmanship to prevail. The Italians, however, had airframes comparable to the S.5, and more powerful, more compact engines in two of their aircraft gave them a definite speed advantage.

Fortunately for the British, Jacques Schneider's original intention that the races should test speed over a significant distance had survived successive revisions of the rules. Fiat had pushed the engine too far, and the spectators saw one of the two M.52s descend trailing smoke into the Laguno within seconds of crossing the starting line. Mario de Bernardi, winner of the 1926 race, lasted rather longer until, as C. G. Grey of *The Aeroplane* put it, "a connecting rod came out of the crankcase for a breath of fresh air".

Not only did the S.5 win the race, but in doing so over a 189nm (350km) course it beat the world's absolute speed record, which had been set by a landplane over a 3km dash. The S.5 averaged 281·65mph; in March 1928 the Italian team managed to make the AS3 hold together long enough to set a new air speed record. It was thanks to Jacques Schneider, who died a 49-year-old pauper in the same year, that the absolute speed record for aircraft (and, for that matter, any man-carrying object) was to be held by seaplanes for more than a decade.

So, in only three years, the competitive response to the Americans' run of success in the Schneider races had left the United States out of the contest. It was noted earlier, however, that there was also an imitative response to the spectacular US assault on the Trophy of 1923. It was to be highly influential in shaping combat aircraft throughout the remainder of the piston-engine era.

In 1920–21, British aircraft manufacturer Richard Fairey (later Sir Richard) was preoccupied by the need for a low-drag engine installation. His company had compared the wartime Fokker D.V fighter with the British S.E.5, and had concluded that the German fighter's speed advantage was partly attributable to its cleaner engine installation. In the immediate post-war years there was no British engine in sight which could offer much aerodynamic improvement over the flat-nosed S.E.5 powerplant. Bristol and Armstrong Siddeley were building air-cooled radials, while Rolls-Royce was mainly preoccupied with rebuilding its car business: the new Condor aero-engine of 1923 turned out to be a scaled-up descendant of the wartime Eagle, still using separate cylinders and water jackets, with a weight and bulk in generous proportion to its output. Napier, after considering the design of a new monobloc V-12 engine, decided that the market was not worth the development cost and concentrated on cleaning up the external shape of the Lion.

When the American team brought the Curtiss racers to Cowes, Fairey realised that his search for a new engine was over. The aircraft manufacturer obtained a licence to build the D-12 and the Reed high-speed metal propeller, and designed the Fox light bomber around the Curtiss engine. Apart from the experimental Hawker Hornbill, designed around the clumsy and indifferent Condor, no RAF fighter could overhaul the Fox. The Air Ministry was impressed with the aircraft, but not with the idea of setting up a fifth company in the aero-engine business, producing copies of an American engine to boot. Accordingly, a D-12 was shipped to Rolls-Royce; at that time the Derby company was working on an ambitious and complex supercharged X-16 engine, the Eagle XVI. The Air Ministry intimated that something along the lines of the D-12 might be more acceptable, and Rolls-Royce began design of a monobloc, narrow-angle V-12 with an integral supercharger geared to the tail of the crankshaft. Work started in July 1925, and the first Falcon X prototype ran in the following year. In service the new V-12 was usually known as the F-type engine; rather later it was given the name Kestrel.

The Falcon X supercharger gave more trouble than was expected, and in 1927 the Air Ministry released the RAE's supercharger genius, J. E. Ellor, to Rolls-Royce. By May of the following year the supercharged Kestrel was in production. Ellor and other Rolls-Royce engineers went on to devise a forward-facing intake for the supercharger, so that

For the next Supermarine Schneider contender, the S.5, Mitchell chose a more conservative braced wing. By this time the Lion was developing more than twice as much power as it had produced for the Sea Lion five years before *(Flight International)*

Rolls-Royce's homage to the achievements of the D-12 was the Kestrel, sharing the same basic layout but adding for the first time an efficient supercharger. The blown Kestrel first entered service on the Hawker Fury, last of the biplane fighters

the forward speed of the aircraft added to the compression in the blower, and thus to the power of the engine, and the first automatic boost control – a governor which ensured that the pilot did not have to worry about allowing the engine and blower to overspeed and create damaging overpressures in the induction system. The first, unsupercharged, F engines went into Fox day bombers, and early examples of the very similar Hawker Hart; supercharged engines were fitted to the first RAF fighters designed to catch and destroy such aircraft. Specification F.20/27 called for an aircraft to overtake a target flying overhead at 20,000ft and 150mph. It was contemporary with the specification that produced the Bristol Bulldog, designed for night operations against slower targets. The specification was noteworthy in that it attracted a monoplane entry from de Havilland; it also attracted interest from all the engine manufacturers. The de Havilland D.H.77 monoplane was powered by a highly unconventional air-cooled 16-cylinder engine, built by Napier in collaboration with de Havilland engine designer Frank Halford. Hawker chief designer Sydney Camm (later Sir Sydney) offered a more conventional aircraft: although originally built with a production version of the Bristol Mercury racing engine, it was the basis for the private-venture Hawker Hornet with the new Rolls-Royce V-12. It was the last-named combination which was ordered, as the Fury and the Kestrel, by the RAF.

Although the Kestrel was undoubtedly inspired by the D-12, its development and its production history started where the D-12 and its Curtiss derivatives – the V-1400 and V-1550 – had come to a virtual halt. For a number of reasons, the development of the water-cooled engine stagnated in the USA in the late 1920s. Possibly the prime cause was the fact that two major sections of the market preferred air-cooled engines: the expanding airmail and airline operators were more concerned with reliability than with maximum speed, while the US Navy's pilots were understandably wary of a breed of engine which had, in their view, one more way to fail. Another important factor was that the

development of air-cooled radial engines quickly matched, and in some respects surpassed, the development of liquid-cooled engines. It should also be remembered that the performance gap between the two types of engine was not as great in practical terms as the Schneider contests would suggest. The high speeds of the racing aircraft were attained using low-drag skin radiators, which were neither reliable nor safe for regular use. (The German aircraft industry was the only one to persist with surface cooling for service aircraft, with results which were generally beneficial to Germany's enemies.) Until the appearance of high-powered, low-frontal-area engine installations like that of the Fury, the radial and the in-line were very closely matched.

America's highly successful radial engines had emerged from a commercial/political mêlée between 1922 and 1925. In 1922, the Wright company had been persuaded by the US Navy to buy the Lawrance Aircraft Company, which was designing small but promising radials: the Navy's idea was that Wright's resources would blend well with the smaller company's ideas. Then, in late 1924, Wright president Frederick Rentschler resigned and (once again, with encouragement from the Navy) set up a new aero-engine operation for a machine-tool company in Hartford, Connecticut. The engineering company's new division was officially incorporated in July 1925 as the Pratt & Whitney Aircraft Company. Wright and Pratt & Whitney, with their strong preference for radials, were to dominate the US engine scene until the flowering of the jet age.

If America's engine manufacturers had temporarily abandoned the pursuit of speed for its own sake by 1928, the loss of their efforts was more than balanced by the energy which Britain and Italy poured into preparations for a grudge race in 1929; and if the engine builders of the New World were conservative, their primary clients on the sunny West Coast were not. A number of US aircraft builders, in fact, were already working on designs which would do for the airframe scene what the D-12 had done for the aero-engine.

300-450mph

Monoplanes and monster engines

300–450mph

Monoplanes and monster engines

BY the time that the Schneider race following Britain's 1927 victory took place in 1929, the contest had departed from the mainstream of airframe development. The interval between races had been established at two years in recognition of the enormous efforts which Britain and Italy were prepared to expend on winning the Trophy, but designs were already in hand in the United States which made the racer airframes look antiquated. There was an enormous gap between the engines and the airframes of the Schneider aircraft: what had to happen, and what did happen within a very few years, was that the engine technology of the racer was wedded to the new ideas in airframe design, bringing about an astonishingly abrupt change in the design of every category of aircraft. In 1930, most military and civil aircraft in production could have chugged above the fields of Flanders a decade and a half before and scarcely attracted a second glance. Within another five years, the aeroplane had changed beyond all recognition.

What the aircraft industry experienced during the years 1927–32 was a form of "synergy" as technical innovations generated the need to apply other innovations in an accelerating, expanding process. Hardly any of the new ideas in aircraft design were conceived after 1927, but their use together was new. All of them were associated, directly or indirectly, with the quest for greater speed.

The oldest of these "new" ideas was monoplane design. As we have seen in earlier chapters, the monoplane antedated the biplane, and was eclipsed because there were fewer uncertainties involved in the design of a safe and lightweight biplane wing. In the late 1920s, the monoplane was still held in some suspicion, due to a number of bad experiences. The US Army Air Service lost two experimental monoplanes to wing failure in 1922. In Britain, a handful of monoplane fighters had been tested, including the Vickers Vireo carrier fighter, tested in 1928 and successfully operated from HMS *Furious*; but the Vireo had a thick, awkward wing design, and after the type had been rejected for service use it was found that the flow interaction around the wing's junction with the slab-sided fuselage was to blame for its poor stalling behaviour. Two other monoplane designs for the RAF were the de Havilland D.H.77, which competed unsuccessfully against the forerunner of the Fury, and the Westland Wizard, a "parasol" monoplane in which some of the structural uncertainties were reduced by fitting bracing struts. The advantages were reduced along with the uncertainties, and like all other British monoplane

fighters of its time the Wizard never progressed beyond the prototype stage.

There was a basic and simple reason for the failure of these pre-1930 monoplanes to attract military interest. Fighter requirements were framed with the biplane in mind, and in particular drew heavily on combat experience in the First World War. A low landing speed was compulsory. High initial rate of climb was a premium quality, because it ensured the advantage of the dive out of the sun in combat. These attitudes underlay the requirement for a Fury replacement, F.7/30, which the RAF issued at the turn of the decade. The various stipulations of F.7/30 ensured that it could not be met satisfactorily by a monoplane.

Also, the development of a successful high-speed monoplane for practical use depended on the availability of a suitable aerodynamic and structural design. The excellent commercial monoplanes of the 1920s–Fokkers, the closely related Fords and the Junkers–had wings so thick that they were only a little faster than contemporary biplanes. The Schneider racers, by contrast, had short, heavily braced wings of thin section, developing poor lift at low speed. This was the challenge facing the designers of the early monoplanes.

All-metal airframes went back a long way by the late 1920s, but like the monoplane they had not yet been brought to a stage of development where their overwhelming advantages stood out. The use of aluminium alloys on a large scale had been pioneered by the Zeppelin company; from the LZ.1 of 1900 onwards, all Count Zeppelin's aerial monsters were composed of a light alloy frame, skinned in fabric and enclosing a series of gasbags. The Staaken aeroplane company, associated with Zeppelin, was a pioneer of all-metal construction for aerodynes; the leader in the field, however, was Professor Hugo Junkers. His first all-metal aircraft, the J.1 ground-attack type, flew in 1917; although it was a biplane, its structural philosophy was that of a monoplane because there were no interplane struts or bracing wires. Multiple wing spars were joined by internal bracing tubes, lugged together, and added strength was given by 0.38mm alloy sheet, which was corrugated for stiffness.

The J.1 was the ancestor of all Junkers' later metal monoplanes, throughout the company's single-engined airliners of the 1920s to the Luftwaffe's workhorse transport of the 1939–45 conflict, the much loved and virtually indestructible Ju52 *Tante Ju*. The Junkers construction method inspired Ford's designers in the USA and Michel Wibault in France–the unfortunate Vickers Vireo was based on Wibault's designs–with the result that a great proportion of early metal aeroplanes featured similar corrugated skins. Structurally such skins were very convenient and saved a lot of weight: aerodynamically they were a disaster.

During the 1920s a number of prototypes appeared with a different kind of load-bearing (or "stressed") skin, which presented a smooth outer surface to the passing air. The experimental Staaken E.4/20 monoplane airliner, the Short Silver Streak and the Supermarine S.5 were examples. Although methods differed in detail, the principle was similar. Aluminium sheet was excellent in terms of a ratio of tensile strength to weight, but on its own would buckle instead of resisting any compression stress. The thin aluminium skin was therefore reinforced with channel sections to prevent buckling, and the whole skin was held in shape by transverse ribs. The internal structure was thus a "semi-monocoque". In the 1920s, however, this method of construction was applied only to the fuselage, where the designer could make use of abundant internal space. The wing presented more difficult problems: chiefly, that of achieving a rigid and light structure, without corrugation or an excessively thick wing section.

Construction on these lines was entirely different from the techniques used in early "all-metal" fighters such as the Fury and the Bulldog. In those aircraft, metal was used as a substitute for wood, in a traditional fabric-covered airframe. For a considerable time Hawker was to resist the change to semi-monocoque fuselage construction, preferring to retain the Fury's steel-tube-and-fabric design.

As more and more all-metal monoplanes appeared on the drawing boards, the prospects of sustained high speeds became clearer to the designers. They began to look again at a number of drag-reducing ideas which had been current for some time, but had been judged unnecessary. Air resistance increases much faster than speed: a gadget that reduces air resistance, but weighs too much or is too complicated to be worthwhile at a certain speed, may be highly desirable at a slightly higher speed. In the early 1930s a whole group of ideas fell into this category. The most immediately obvious of these was the retractable landing gear. It was an old idea; in 1876 Alphonse Pénaud had proposed retractable gear for one of his aircraft designs, as a means of allowing it to fly from land or water. Retractable wheels had been used for the same reason on a number of amphibious aircraft, and the Dayton-Wright racer of 1920 was the first to make use of such a device in the interests of higher speed. Enclosed canopies came into the same category. They had been used on a number of aircraft since the abortive attempt to persuade pilots to fly the S.E.4 with such a device; these were mostly commercial types, and the aim was comfort as much as aerodynamic improvement. It was some time before designers managed to demonstrate that

Retractable undercarriage, enclosed cockpit, a small unbraced wing with flaps on the leading and trailing edges: all these features became popular during the 1930s, but the astonishing Dayton-Wright RB-1 racer was flown in 1921 (*Smithsonian Institution*)

29

enclosed canopies were now necessary for high-speed fighter aircraft.

Designers were also paying more attention to the relationship between low-speed and high-speed performance, and in particular to the ways in which technology to improve low-speed handling and control could, indirectly, improve matters at the high end of the scale. Both the S.E.4 and the Dayton-Wright racer had been designed with relatively small wings for low drag and high speed, and both used wing flaps to increase the camber and hence the lift of the wing at take-off and landing. Thereafter, the industry at large tended to lose sight of this application of wing flaps. Dr Gustav Lachmann of Handley Page developed powerful systems of full-span leading-edge slots and trailing-edge flaps, in some cases with ailerons that drooped at low speeds to extend the flaps over the entire wing. However, these were usually considered as aids to safer flying, rather than as a means to reduce the size of the wing; one of the few high-performance aircraft of the 1920s to use such devices was Handley Page's own H.P.21 fighter, a cantilever wooden monoplane with full-span slots and full-span trailing-edge flap/ailerons. Its powerplant—somewhat archaic in an aircraft flown in September 1923—was a Bentley BR.2 rotary. The H.P.21 was actually ordered by the US Navy—no doubt the service was attracted by its

44mph landing speed—but the contract for 27 HPS-1s was cancelled before the aircraft were delivered. It was to be more than ten years before some of Handley Page's ideas were to take shape on a fighter aircraft.

The first step on the way to incorporating all the foregoing features in a single aircraft took place in 1926, when Allan Lockheed and Jack Northrop started work on a new light transport aircraft. Although Northrop's dream was to build tailless flying wing aircraft in metal, he turned his hand with great success to the wooden construction favoured by Lockheed. Aircraft and boat manufacturers had recently discovered casein glues, which had been used now and again since the days of the Pharaohs: these made possible far stronger and more weatherproof wooden structures. The Vega had a cantilever wing and a plywood fuselage formed on a concrete mould, and despite the jutting cylinders of its radial engine looked remarkably clean at the time of its first flight in July 1927. By the following year the cylinders had disappeared. They were concealed beneath a new type of cowling developed by the National Advisory Committee for Aeronautics (NACA). NACA had been precisely what its full name implied until 1926, when Congress expanded its brief to include applied research along similar lines to Britain's RAE. The NACA

Lockheed's Vega was the first of the clean, efficient new monoplanes which emerged from the US West Coast in the late 1920s. A year after appearing in this form the Vega was fitted with a streamlined NACA cowling and won the main prize at the National Air Races in Los Angeles *(Smithsonian Institution)*

cowling for air-cooled engines was one of the first of many technical advances to emerge from the organisation. Consisting of a large-diameter tube encircling the engine, the NACA cowling eliminated the turbulent interaction between the air that was needed to cool the engine and the rest of the free stream; the cooling air was returned to the free stream through a carefully shaped annulus at the back of the cowling. Tested on a Curtiss AT-5A trainer in 1928, the new cowling increased the speed of the aircraft from 118mph to 137mph.

NACA presented the new idea to the Army Air Corps (the Army Air Service had been so renamed in 1926), but fighter pilots reacted against it, insisting that their downward and forward view between the cylinders was essential for safe flight. Like a number of British aircraft, therefore, the next few USAAC fighters used an alternative low-drag cowling developed by H. Townend of the National Physical Laboratory; the Townend Ring was a short-chord cowl which did a similar job, was not quite as efficient but left the view between the cylinders unobstructed.

The NACA cowling was eagerly adopted by the commercial side of the business, however, and one of the first aircraft to use it was a Lockheed Vega which, in 1928, won the main prize at the National Air Races in Los Angeles. By this time the US armed services were reduced to entering standard service types in air races; the Vega beat the fighter biplanes easily, and in the following year, when the National Air Races moved to Cleveland, a little 450hp monoplane built by Travel Air and known only as the "Mystery Ship" inflicted an even more humiliating defeat.

The Vega's victory in 1928 sent some racegoers home in a thoughtful mood. The Boeing company of Seattle was preoccupied with the mail business, because its main civil market was with a number of associated companies in mail-carrying. It was also interested in fighters, because it was producing the basically similar F4B and P-12 fighters for the US Navy and the Air Corps respectively. The sort of technology embodied in the Vega could make its mark in both areas. The company was already working on the XP-9 fighter for the USAAC, with a braced monoplane wing and stressed-skin metal fuselage, and this may have influenced Boeing to make the design of its next mailplane a somewhat bigger gamble.

The new aircraft was the Model 200 Monomail, one of the most successful failures in aviation history. Flown in May 1930, it was the first design to combine an all-metal structure, a cantilever monoplane wing, a low-drag engine cowling and a retractable undercarriage. Its powerplant, how-

A step beyond the Vega was Boeing's all-metal, retractable-gear Monomail, seen here in an early form with an uncowled engine. Like Boeing's earlier commercial aircraft, it was designed for use by airlines which made most of their money from mail, and the passenger cabin was accordingly small and cramped (*Smithsonian Institution*)

ever, was the source of a disappointing performance in the air. The Monomail's propeller had blades which could be pivoted at the root to change the pitch before flight. If the propeller was set in "fine pitch", with the blades at a small angle to the plane of rotation, the Monomail took off readily. However, as its speed increased, the blades would strike the faster-moving air stream at a steadily shallower angle, and become steadily less efficient until the Monomail settled at a mediocre cruising speed. Alternatively, the ground crew could set the blades in "coarse pitch", with the blades at a larger angle to the plane of rotation. At rest or at low speed, however, this meant that the blades were striking the air almost sideways, and even in cool weather at low altitudes it was all that the Pratt & Whitney R-1830 could do to turn the propeller fast enough to get airborne. The Monomail was, in any case, a rather large aircraft for a 575hp engine, and the propeller problems ensured that it never saw service.

Boeing hoped that the next commercial aircraft on its drawing boards, with two engines and about 50 per cent more weight, would do better; but it does not take an advanced knowledge of economic history to know that 1930 was not the time to be selling new, advanced and expensive commercial aircraft in the United States. Instead, Boeing mated a bomber fuselage with the wings of the proposed commercial twin, and built a prototype as a private venture.

The time was ripe for such an aircraft. If the technical advances of the 1920s had been adopted with limited enthusiasm in fighter design offices, they had bypassed large offensive aircraft completely. The USAAC bomber squadrons were equipped with 1920-technology drag factories such as the Keystone B-4 and B-6, which could attain 121mph flat out and cruised at just over 100mph. Unlike the fighter squadrons, the bomber units did not have to worry about rate of climb or manoeuvrability.

Early tests of the new Boeing, the XB-901, were successful, and in August 1931 the USAAC agreed to buy the prototype and six more examples. The first aircraft, re-engined with mildly supercharged R-1830-11 engines, could cruise at 165mph. This may not seem much, but it was no less than 60 per cent faster than the miserable Keystone. The USAAC was at last motivated to defy the politicians from Pennsylvania who had ensured that the Keystone stayed in production, and it terminated production of the biplane in the following year.

The USAAC did not fail to notice that it was testing a bomber which could cruise at very nearly the maximum speed of its standard fighters, and in the same year that it decided to test the XB-901 – then redesignated B-9 – the service agreed to test a monoplane fighter from Boeing. Flown in 1932 as the XP-26, the fighter was a compromise between the manufacturer's advanced technology and the customer's conservatism: the undercarriage was

32

fixed, and the AAC insisted on bracing wires for the wing, these extending from the large undercarriage fairings and the upper fuselage. Insistence on a good downward view for the pilot meant that the cockpit was installed at the highest point of the fuselage top line, and a large roll-over pylon was installed to protect the pilot's neck. As a result, the P-26 as it eventually entered production was not the fastest of monoplanes; in fact, it was only marginally quicker than the contemporary late-production Fury biplane. With the French Dewoitine D.500, however, it constituted the first generation of production monoplane fighters, and it was also the first production fighter to be fitted with flaps.

Boeing's rivals further down the West Coast had not been idle while the B-9 and P-26 were under development, but their ideas took a little longer to come to fruition. Lockheed had developed the high-wing Vega into the low-wing Sirius, with two seats in tandem under one of the first clear cockpit canopies; designed for route-proving and survey flights, the Sirius was used by Charles and Ann Lindbergh as well as other pioneers. In 1930 Lockheed flew the Altair, a development of the Sirius with a hand-operated retractable landing gear, and from this developed the XP-900 prototype fighter. Apart from its wooden airframe the XP-900 was an early embodiment of many features which were to become familiar: a low-wing cantilever monoplane with an enclosed cockpit, powered by an in-line liquid-cooled engine (one of the last descendants of the Curtiss D-12), and fitted with a neat and simple landing gear which retracted upwards and inwards into the wing. It was only a little faster than Lockheed's Orion seven-passenger transport, which could cruise at 200mph on 450hp and was sold to a number of airlines for high-speed express services. In 1931, however, Lockheed's owner, Detroit Aircraft, fell victim to the Depression, and the Lockheed line of wooden high-speed aircraft came to an end. For the time being, metal construction became universal for such machines. Jack Northrop, co-designer of the Vega, had left the company to pursue his own ideas for all-metal, all-wing aircraft, and built a number of prototype aircraft from 1929 onwards in which the cantilever monoplane wings were constructed of shaped and riveted alloy sheet sub-assemblies, forming spars and ribs, built up into a structure that functioned as a series of torsion boxes. Easy to build, light and rigid, the Northrop "multicellular" system was easily the most advanced structural technique of the time. Its immense strength was inadvertently demonstrated when one of Northrop's prototypes crashed. Northrop wanted to scrap the wreckage, and borrowed a small steam-roller from a nearby road gang; one of the crew drove the roller backwards and forwards over the wing several times without the slightest effect.

In the same year, a former associate of Hugo

Junkers became director of the newly founded Guggenheim Aeronautical Laboratory at the California Institute of Technology in Stanford. Theodore von Karman was a friend and colleague of many of the new young designers in the USA, and was soon working with Northrop on the problems of metal cantilever wings, and in particular on the post-buckling qualities of properly designed aluminium structures. Internationally, there was rapidly growing confidence in metal, and in the feasibility of designing thin-section, high-strength cantilever wings.

The difficulties of propeller design remained a limitation on practical aircraft; unlike racers, they could not use brute force to overcome the low-speed inefficiency of a coarse-pitch propeller. The USAAC had experimented with a two-speed reduction gear, but gave up after the synchromesh box was wrecked by an attempted change in a power dive. The problem had not been finally solved when, in mid-1932, United Air Lines gave Boeing a long-awaited order for the commercial twin which the company had designed in 1930. However, Frank Caldwell of Hamilton Standard, one of the principal US propeller manufacturers, was in the final stages of designing a propeller which incorporated a mechanism to change the pitch of the blades in flight. This had long been recognised as a solution to the problem, at least in theory; the difficulty lay in developing a pitch-change system which would be effective and reliable. If the system were to collapse, the blades might be free to pivot into full fine pitch. In that condition, they would not merely produce no power; they would generate so much drag that the aeroplane would almost certainly crash. By mid-1933, however, when Boeing's new Model 247 was demonstrating the same disappointing hot-weather, high-altitude performance as had marred the Monomail, Hamilton Standard's new variable-pitch propeller was ready for production. It was adopted for the improved Boeing 247D, which began to reach customer airlines in 1934. In the same year, Douglas started production of the DC-2, using Jack Northrop's multicellular wing; and virtually standard examples of the two American airliners came within a hair of winning the London–Melbourne air race of that year. They were prevented from doing so only by Geoffrey de Havilland's rapid development of a specialised long-distance racing aircraft: this, the D.H.88 Comet, was the first British aircraft of any kind to fly with variable-pitch propellers, even as they were being used on regular passenger flights in the USA.

The result of the MacRobertson air race – a narrow victory by specialised racer over standard airliner – was a rude shock to Europe, which had imagined itself the leader in high-speed aircraft. This was true, up to a point; even by 1934, no US pilot had matched the 310mph-plus speed marks set by the 1927 Schneider racers and lived to claim an official record. (The last qualification is necessary:

Lowell Bayles was believed to have attained 314mph in December 1931, on one record run in the vicious Gee Bee Model Z, but was killed on the second run over the course.) The British and Italians, meanwhile, had further raised the record in the course of the 1929 and 1931 Schneider contests.

The European racers, however, had achieved their ends by sheer brute power. The result was that Europe, and particularly the British, knew a lot about very powerful, relatively light and very low-drag engines, but had not built practical airframes to exploit them. Once they managed to do this, however, the Old World quickly came to dominate the high-performance scene; and within a very few years a great deal more than the ownership of a silver cup was to depend on aircraft and engine performance.

The single technical development for which the last two Schneider races are best remembered arose from a decision by Major G. P. Bulman, assistant director for engine development at the British Air Ministry: the 12-year-old Lion, however modified, could not be expected to hold off the determined and well-funded Italians' attempt to wrest back the Trophy in 1929. In late 1928 Bulman took on the job of trying to persuade Rolls-Royce not only to build a racing engine – Rolls-Royce, of course, had never engaged in any sort of competition – but to do it in less than a year. Sir Henry Royce assented after Bulman and one of Supermarine's directors made a personal appeal to his patriotism. Work on the new engine, known simply as the R, began in November 1928.

Italy, meanwhile, was preparing no fewer than four aircraft types from which its team of three individual aircraft could be selected. Easily the oddest was the Piaggio P.7c: the design aimed at dispensing with drag-producing floats, so the fuselage was made watertight. A water propeller and hydrofoils were fitted (as on the Bristol X.2 of 1912) to propel the aircraft in its first stage of take-off and lift it clear of the water. The P.7c was a total failure, because problems with the drive to the water propeller prevented any "transitions" to foil-borne running. Another unconventional design was the Savoia S.65, with two engines installed fore and aft of the cockpit and driving tractor and pusher propellers; the tail was carried on twin slim booms. The S.65 was not sufficiently developed to compete in the race, and crashed in January 1930 during an attempt on the world's air speed record. The other Italian aircraft were conventional tractor monoplanes: the Macchi M.67 was based on the preceding M.52, but fitted with a heavier, more powerful and more reliable W-18 engine (three banks of six cylinders) from Isotta-Fraschini. Fiat, excluded from Macchi's aircraft and still in disgrace from the AS3 fiasco in 1927, developed its own C29 racer to use the all-new AS5 engine. Only the M.67 was ready in time for the race.

Frantic efforts on the part of Rolls-Royce and Supermarine produced the British team's great hope on time. Rolls-Royce had used an existing engine as the basis for the R – the Buzzard, a scaled-up version of the Kestrel with 87 per cent more capacity, which had been developed for use on large aircraft such as flying boats. The basic idea was to feed a Buzzard through a huge two-sided centrifugal blower, then by far the largest ever built, the diameter and hence the power of which would be determined by the diameter of the racer's fuselage, and to get as much power as possible without breaking something.

It was a task that called for simultaneous work in a whole range of engine technologies. Chief designer A. J. Rowledge – who had designed the Lion for Napier, and the Kestrel for Rolls-Royce – was responsible for ensuring that the many moving components of the engine survived unprecedented pressures, temperatures and speeds. James Ellor designed the world's biggest supercharger, also the first to operate successfully with a dual-sided rotor and a ram inlet, and F. R. "Rod" Banks of the Associated Ethyl Company blended the curious concoctions of exotic hydrocarbons which the monster turned into power.

Banks' activities were particularly important, because the big supercharger, operating at ground level, was feeding the R with its air/fuel mixture at higher pressures than had been used before anywhere. Detonation – the ignition of the mixture in the cylinder through compression, before the spark fired or the piston reached the top of the upstroke – was a constant limit to the engine's performance. The science of creating high-performance fuels was in its adolescence in 1929. After the First World War, and the bad experience with US gasoline (see first chapter), a group of British investigators – which included young and rising scientists such as Henry Tizard, David Pye and H. R. Ricardo, all of whom were to make immense contributions later in life – carried out some research on fuel properties and produced some simple and, within limits, effective tests for anti-knock quality. At about the same time, however, the US Army Air Service technical division was finding that while the Liberty engine needed added benzol to run properly on domestic aviation grade fuel, the Curtiss D-12 would run without benzol at similar compression ratios – clearly the engine design had some influence on the question.

A further step to solving the problem came from the motor industry. In 1911, a subsidiary of General Motors which made ignition equipment – Delco – began to investigate complaints that its ignition systems caused knocking. C. F. Kettering and Thomas Midgely of Delco started to identify compounds that worsened or ameliorated the problem; they found that a chemical called tetra-ethyl-lead (TEL) considerably improved the behaviour of an engine if added to fuel in quite small quantities. In the early 1920s this work led to the establishment of the Ethyl Gasoline Corporation, jointly founded by GM and Standard Oil, to sell the product, known popularly if inaccurately as "ethyl" or "lead". The US Navy was the first aviation operator to start using lead; from 1926 until 1933, USN aircraft carried cans of TEL, which could be added to the fuel if knocking occurred. The Army made some use of the product from 1927 onwards.

There was still a need for a means to evaluate the performance of different fuels. A chemist working for the Ethyl Gasoline Corporation – which wanted to set standards for fuel to which its product was added – made the major breakthrough in this area. While looking for a low point on the scale, he investigated a pure hydrocarbon called octane and found that its resistance to knock was very high. There was another hydrocarbon, heptane, which was very similar to octane but had very poor knock resistance. Heptane and octane were adopted as opposite ends of what became known as the octane scale; fuels which fell between them were compared under controlled conditions with various blends of octane and heptane. In March 1930 the Army Air Corps specified 87-octane "Fighting Grade" fuel; the octane system was used universally until the last days of the piston engine, when it began to be replaced, at least for research purposes, with a Performance Number (PN) scale which more accurately reflected the actual performance of the fuel. In terms of PN, the fuel specified in 1930 represented a 40 per cent improvement over some fuels previously used, and in a properly designed engine that improvement could be translated directly into power.

The octane system was not generally applied in Britain at the time of preparation for the 1929 race, and instead Banks relied on the known anti-knock properties of benzol. For the 1929 race, the R consumed a mixture of 78 per cent benzol and 22 per cent Romanian gasoline with a shot of tetra-ethyl-lead. This was not a practical fuel for everyday service use, mainly because benzol has a high freezing point and would be an embarrassment at higher altitudes and lower temperatures; but its use in the R, some time before standard gasolines with similar knock resistance were in use, gave Britain in general and Rolls-Royce in particular a great deal of valuable experience with a powerplant in which supercharging was part of the basic concept.

In May 1929 the R was bench-run at 1,545hp for 15min; two months later, a one-hour run was accomplished at 1,614hp, and in August the target output of 1,800hp was attained for the first time. Meanwhile, the mating of the engine with the new Supermarine airframe was under way. The new S.6 racer was essentially a scaled-up S.5, with metal instead of wooden wings. The surface radiators had now expanded to cover part of the floats, and were further enlarged after early tests. Even the tailfin was transformed into an oil cooler: the hot oil was

A racing seaplane like the Supermarine S.6B was practically a flying radiator for the 2,350 hp Rolls-Royce R engine (1). The complete wing surface (2) consisted of double metal skins, the water flowing through channels (3) formed by a corrugated core. The floats (4) were similarly treated. The header tank (5) and associated filler cap (6) and vent pipe (7) had to be located close to the centre of gravity, and this meant that they were directly ahead of the cockpit, which contained the crucial temperature gauge (8); like some later high-supersonic aircraft, the racers had to be flown according to temperature. Some of the feed pipes (9 to 12) helped in the cooling process. The whole system was energised by an engine-driven water pump (13). A separate and complex system was needed to cool the oil, using an integral tank/radiator (14) in the fin and radiator panels (15) on the fuselage side.

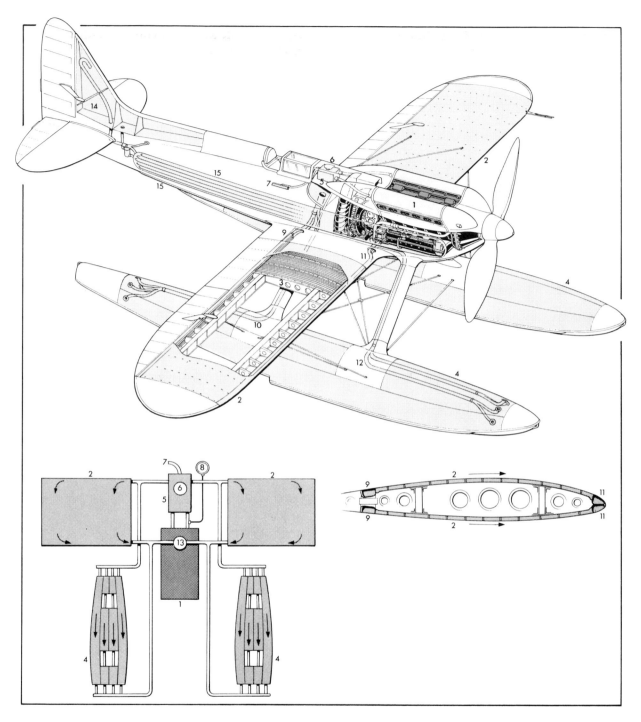

sprayed on to the inside of the metal skin, and dripped down to be collected at the base of the fin.

September 1929: a matter of a few weeks after the engine ran at full power for the first time, the R was rated at 1,900hp for the Schneider Trophy race at Cowes. Even the M.67 was a generous 20mph slower than the S.6 on the laps which one aircraft managed before dropping out with a water leak. The only Italian aircraft to finish was an M.52R, closely comparable to the S.5, and it proved some 40mph slower than the new British aircraft's 328·63mph average. Shortly afterwards, in the order of things which had become traditional, the

Schneider winner attacked and broke the world's air speed record, raising it to 357·7mph.

As noted earlier, the Italian team attempted to beat the S.6's record in January 1930, losing the S.65 and its pilot in the process. Castoldi, the designer of the Macchi racers, had nevertheless been impressed by the logic of the S.65's push-pull layout. It gave the power of two engines for the frontal area of one. It also eliminated the torque reaction of the conventional racers. This characteristic, caused by the equal and opposite reaction of the engine and airframe to the rotation of the propeller, had been getting steadily worse as the power of the racers increased and propellers of ever coarser pitch were fitted to attain higher speeds. With a high-speed propeller fitted, the engine had to push the blades almost sideways through the air while the racer was taxiing or on its take-off run; the immense torque reaction came as the control surfaces were moving barely fast enough to have any effect at all.

The S.65's tandem layout, with the propellers turning in opposite directions, solved the worst of the problems associated with torque, but posed new difficulties in the design of the twin-boom airframe. Castoldi applied push-pull in a new way, which permitted a more conventional airframe. On his instructions Fiat mounted two AS5 engines on a common crankcase, nose-to-nose; naturally, the two crankshafts rotated in opposite directions. Each shaft drove one of a pair of concentric power shafts through reduction gears located between the engines; the concentric shafts ran forwards between the cylinder banks of the front engine. The end of the hollow outer shaft carried one two-blade propeller; the inner shaft protruded from the open end of the outer shaft and carried another propeller. The two counter-rotating propellers were thus driven entirely independently. As far as valve trains, crankshafts and other moving parts went, the

engine was two standard V-12s of a well developed type; in output, however, the new AS6 was a V-24 of unprecedented power, promising 2,500hp initially with potential for a great deal more, and with a relatively small frontal area. The new MC.72 – Castoldi's immense contribution to the Italian racing effort was at last recognised – was designed around the AS6. Essentially it was a low-drag flying radiator on pontoons, its wings covered with flattened copper tubes for cooling water. Like all the later racers it carried fuel in its floats: metal tanks, built into the wooden hulls, were pressurised with compressed air to raise fuel into the fuselage header tank in high-"g" turns. (As fuel was used, a mixture of compressed air and gasoline vapour must have built up in the tanks, turning them into something closely resembling the highly destructive fuel/air explosive weapons of today.)

In 1930, however, it began to seem that all of the Italian preparations might be a waste of time due to lack of competition. Britain's National Government, a coalition formed to rescue the country from depression, had decided that racing was an unjustifiable use of public money. The leader-writers orated and thundered, but Britain's administration remained unmoved by words alone. Hard cash – a

Schneider Trophy races became a contest of engines rather than airframes. The Supermarine S.6 closely resembled the smaller S.5 of 1927 but its cleaner nose housed a powerplant more than twice as potent as the Napier Lion: the supercharged, V-12 Rolls-Royce R

£100,000 donation from Lady Houston – formed the basis for a compromise. With a grudging manner to placate the Members of Parliament from the more depressed areas, the government appeased the Blimps and Bulldogs by releasing the full resources and fleet of the High Speed Flight to assist a privately funded British entry. But by the time agreement was reached, the end of 1930 was approaching fast. The British had nine months to match the MC.72.

There was neither time nor money for a new aircraft or a new engine. The outcome of the contest rested, it appeared, on Rolls-Royce: and many times during the spring and summer of 1931 the engineers at Derby must have rued the day that Sir Henry was talked into the racing business. Supermarine, Rolls-Royce and the High Speed Flight had considered the competition, and decided that 2,300-2,350hp in a modified S.6 would hold off the Italians. By April the snarling of four V-12s shook the walls of the Rolls-Royce test-house: one Kestrel to cool the crankcase, a second to pump exhaust out of the building and a third to blast a 380mph airstream into the ram inlet of the R, bellowing and straining and putting more power into the dynamometer than any single engine had done before.

The Rolls-Royce men might have taken some comfort had they been able to hear the noises from Fiat's test houses, where the AS6 was imitating a duel between a pair of maddened allosauri. The major and fatal weakness of the AS6 stemmed from early wind-tunnel tests of the coupled-engine concept, which had shown that the rear airscrew could absorb more power, and develop more thrust, than the front propeller – the effect was ascribed to the fact that the rear airscrew operated in the accelerated slipstream of the other unit. The AS6 was therefore arranged so that all the power of the forward engine was used to drive the rear airscrew. The rear engine could produce more power than the front airscrew could use, so it was used to drive a single supercharger for both engines, located at the back of the whole AS6 complex.

Efficient and thoughtful, this arrangement nevertheless meant that the whole AS6 had a single system of pipes carrying hot air and fuel vapour from the blower to the cylinders, some of which were about seven feet away. In terms of induction systems, seven feet is a very long distance, and a great many things can happen in a little more than two yards. In the case of the AS6, poor fuel distribution led, at best, to rough running and, at worst, to backfires. A backfire, or explosion in the induction system, when performed by a small car engine, can send law-abiding citizens diving for safety; the AS6, with 51·1 litres and 24 cylinders, was prone to blasts on a Vesuvian scale.

As the Fiat engineers worked on the AS6, Rolls-Royce steadily and painstakingly replaced almost every piece of the 1929 R engine. The two basic changes to the engine had been a higher-ratio gear in the supercharger drive, so that the air entering

Fastest of all the Schneider seaplanes was the Macchi-Castoldi MC.72. It did Italy no good, however, because the 24-cylinder AS.6 engine could not be tamed in time to prevent Britain's victory. The ground crew appear somewhat apprehensive, not without reason *(Smithsonian Institution)*

the cylinders was pressurised to 18lb/in² above atmospheric, and a higher speed limit of 3,200rev/min rather than 2,900rev/min. To withstand operation at higher speeds and pressures, the engine needed new connecting rods; then the crankshaft had to be replaced with a balanced type, and the case modified to accept the higher bearing loads. Bronze and steel components were replaced by aluminium forgings, among the first to be used in a Rolls-Royce engine.

Rolls-Royce also made its first acquaintance with a new type of valve, designed to take a much more active role in preventing heat concentrations in the combustion chamber. This was the cooled valve, containing a liquid which partly filled the hollow valve stem and, splashing upwards and downwards, would transfer heat out of the cylinder. These had been neglected in Britain after some unsuccessful wartime RAE work on mercury- or water-cooled valves, but had been pursued in the USA to the point where practical valves cooled by sodium were in production by the late 1920s. These were an important feature of the 1931-model R.

At the end of August Rolls-Royce completed a 58min run at 2,360hp and, early in September, the full one-hour test was accomplished. About the same time the Italians and the French (who were considered outsiders) requested a postponement of the race. The French had lost one aircraft with its pilot; the Italian team was confident that it had a race winner in the MC.72, but the problems of the AS6 haunted them still. Capt Giovanni Monti had been killed after an engine failure in August, and another pilot died in the wreck of an MC.72 in September. Nevertheless, the British Royal Aero Club refused to consider delaying the race.

This decision has been contrasted with the sportsmanlike behaviour of the USA in 1924, when in similar circumstances the Navy team declined to claim a flyover victory; but in 1931 there was a great deal more at stake. The British team, having been forced to start late, had an aircraft which they believed to be a winner, despite the far greater resources committed to the competition. Less than a year before it had seemed that there would be no British entry at all. To delay the race by the months necessary to put the AS6 right would have been an act of sportsmanship indeed, but it would have been a difficult gesture to explain to the deafened engineers at Rolls-Royce, or to the exhaust-roasted, wind-blinded and oil-soaked pilots of the High Speed Flight.

By September the problems of the R had been reduced to manageable proportions: it had been weaned out of an apparently insatiable thirst for lubricant, and Rod Banks had brewed for it a 92-octane tipple of Californian gasoline, benzol and methanol, with the customary dash of lead. The methanol (methyl alcohol), constituting ten per cent of the fuel, was particularly significant. Although it produced less power for a given volume or weight than gasoline, it had the valuable property of requiring a large amount of energy to vaporise it. Methanol used as fuel therefore draws heat out of the charge of fuel and air in the manifold, tending to suppress knocking and allowing the engine to be pushed harder. In the case of the R, as in all subsequent highly supercharged engines, this meant that the opening of the throttle which controlled the airflow to the supercharger (and hence the pressure of the air in the induction system) had to be adjusted carefully to avoid damage to the engine.

The main modifications to the S.6 for the 1931 race concerned the cooling system, which had to do a bigger job more effectively. A significant modification was that the air vents in the cooling system were redesigned to face forwards, so that ram air would slightly pressurise the coolant and, in theory, slightly increase its boiling point. Nevertheless, on the S.6B – as the re-engined and modified aircraft was designated – the instrument which the pilot had to observe most vigilantly at high speed was the temperature gauge.

Race day, after all these preparations, was a little anti-climactic: there was no race, just a 340mph circuit of the course which barely tested the new engine. But with a new methanol-based fuel in the tanks the S.6B put the air speed record above 400mph for the first time – on 2,800hp from the R engine.

Despite the British victory – and the end of the Trophy era – Italy continued to test the surviving MC.72 for more than three years, and with some assistance from Rod Banks they managed to cure the induction problems. Eventually, by October 1934, the AS6 was developing 3,100hp, an output which was to remain unsurpassed for more than a decade. In the hands of Francesco Agello, the MC.72 set a new world's air speed record at 440·68mph. At the time of writing it remains the absolute speed record for seaplanes.

The Schneider racers were freaks, beyond any doubt, and the world will not see such aircraft again, but their influence on the world of aviation was considerable. In Europe they were a very persuasive advertisement for the merits of the supercharged V-12 engine, and the advantages of liquid-cooled, inline powerplants for high-speed aircraft. The development of several such powerplants in all the European aircraft-producing nations in the early 1930s cannot be ascribed entirely to coincidence.

While Agello and the rest of the Italian team were preparing the MC.72 for its last record flight, the results of the London–Australia race were telegraphed home to London. As related above, the performance of the American contenders came as something of a shock. The ultimate aircraft, it seemed, would be born of the union of an American-type airframe and a high-powered European-style engine, driving an American-style propeller. With international tension steadily increasing, this union was not to be delayed very long.

450-500mph

The great piston fighters

450–500mph

The great piston fighters

THE reign of the piston-engined monoplane fighters was as short in relation to aviation history as the lordship of the dinosaurs was to the animal kingdom. Just under nine years passed between the first flight of the first of the class and the event which marked the end of their empire: the use in action of a new and far more advanced breed of aircraft. That just-short-of-a-decade coincided with a war which pushed total production of fighter aircraft well into six figures and spurred the application of a whole range of new technical ideas. That accounts in part for the attention which the few surviving aircraft of the era can draw wherever they go. The bulk of their appeal, though, lies in the unmistakable, inimitable vocabulary of barks, snarls and howls that mark the progress of a thousand-horsepower-plus supercharged piston engine.

Mid-1933 is the point where the history of the great piston fighters can be said to start. The P-26 and the Fury were probably the world's best fighters at the time; the Royal Air Force, not yet committed to expansion, had decided to follow the Fury with another biplane. For the Bayerische Flugzeugwerke (BFW) of Augsburg, however, there was little hope of government orders in the near future. The company was small, with 82 employees. Worse from the viewpoint of future business was, first, that Adolf Hitler's newly appointed deputy commissioner for aviation was Erhard Milch; that, secondly, BFW's chief designer was Professor Willy Messerschmitt; and thirdly, that Milch and Messerschmitt disliked each other with Wagnerian intensity.

BFW had sufficient resources to stay in business without government orders, so Messerschmitt and his staff were able and free to pursue their own ideas. In late 1933, Messerschmitt decided to build a high-performance cabin monoplane, the M-37, to compete in an international aviation performance contest in the following year. The M-37 was a very ambitious design, of all-metal construction with new-type countersunk "flush" rivets and an outward-retracting landing gear. (Germany's first low-wing, retractable-gear monoplane, the officially sponsored He70, had flown only a few months earlier.) The wing of the M-37 was perhaps its most interesting feature: tiny by the standards of the day, it was fitted with automatic slots and full-span trailing-edge flaps and had no ailerons. Small surfaces called "spoilers" extended from the surface of the wing, ahead of the flaps, to provide lateral control.

Before the contest Messerschmitt had time to recognise that he had been too ambitious in seeking to dispense with ailerons, and installed basically conventional lateral control surfaces so that they would droop slightly with the flaps. Even so, the M-37 – redesignated Bf108 following a little thawing of the official attitude towards the BFW – acquitted itself well in competition. Long before it had started on its own distinguished career (the Bf108 was the inspiration for a number of post-war light aircraft) it had been followed on the BFW drawing boards by an aircraft destined for far greater fame.

This was the Bf109, designed to meet an official Luftwaffe requirement for a new interceptor and tactical support fighter. It was closely related to the Bf108 in its structural concepts and aerodynamic design, but was slightly bigger and designed to accept either of two high-powered, supercharged 12-cylinder engines then under development by Junkers and Daimler-Benz. By the late summer of 1935, when the Bf109 *Versuchs* 1 (first test aircraft) was completed, neither of the new German engines was ready to fly. Accordingly, the first flight was made with a British Rolls-Royce Kestrel V, at that time the best engine that could be obtained on the open market.

The year 1934 was important in the evolution of the fighter aircraft. In Britain and the United States as well as in Germany it was realised that the new high-speed commercial aircraft of the day were the new high-speed bombers of the future. If the fighter pilots were to have the slightest chance of overhauling their prey, they would have to forget some of the favourable biplane characteristics which they had demanded on early monoplanes such as the P-26.

Britain's equivalents of the P-26 had been aircraft like the Supermarine 224, designed to the F.7/30 specification. Compromised by landing-speed and visibility requirements, the 224 was not outstandingly superior to the conservative biplanes which were entered in the competition. Its powerplant was also a problem. The Rolls-Royce Goshawk engine, similar in size to the Kestrel, was cooled by a new system: the water in the engine was deliberately allowed to boil, and was ducted to condensers in the wing leading edge. The Goshawk proved highly temperamental, but it did promise a solution to a dilemma which faced fighter designers such as Reginald Mitchell. The inline engine offered a basically well streamlined installation, but with a conventional radiator its margin of superiority over the radial, now fitted with a low-drag cowling, was small. Accordingly, Mitchell persisted with the Goshawk for the new Type 300, the 1934 design which represented the impact of changed operational requirements on the 224.

There were elements of the He70 in the new aircraft, with its elliptical wing planform and outward-retracting main gear, but Mitchell's great achievement was in designing a wing of large area and span – needed for take-off, turn and climb performance – with a very thin section, necessary in

pursuit of speed. This was to be one of the chief elements in the making of the Type 300 Spitfire.

The other main ingredient of the Spitfire's success, its powerplant, was to take shape slightly later than the airframe, in the course of 1934 and 1935. The dissatisfaction with the Goshawk had not evaporated as well as its cooling water, and Rolls-Royce was alert to possible ways of cutting down the drag of a more conventional cooling system. In 1933 the company had become the first engine manufacturer to set up its own flight-test department, in which high-performance aircraft could test the latest engine modifications with the engine manufacturer's own pilots in charge.

One of the first long-standing myths to be exploded by Rolls-Royce's work was that the best place for the radiator was in the full blast of the slipstream, where the radiator itself could be made as small as possible. Experimental work proved the opposite. What was important was not the outside drag of the radiator so much as the resistance of the air passing through it. This could best be reduced by slowing down the flow through the radiator, installed in a specially designed duct. Airflow could be controlled using a simple flap at the exit from the duct, while the radiator and duct could be recessed into the aircraft. In 1935 F. W. Meredith of the RAE

added to this work by demonstrating that duct design could take account of the heat transferred to the airflow by the radiator, so that the heated and expanded air actually accelerated as it left the duct and generated a small but measurable amount of thrust. Thanks to this work, the big flat radiators that had disfigured many liquid-cooled engine installations vanished from the Spitfire and later Bf109s, to be replaced by narrow aerodynamic ducts. Another of Meredith's ideas was even simpler: the ejector exhaust stub, which turned the waste energy of the exhaust into forward thrust.

All this meant that the Supermarine Type 300 could be made to work very well with a conventionally cooled engine. Rolls-Royce had been working on such a powerplant, stemming from a scaled-up Kestrel, since mid-1932; there was no official requirement for it when it made its first run in October 1933, so it was known as the PV-12, indicating that it was a private-venture 12-cylinder engine. Towards the end of 1934 the PV-12 replaced the Goshawk in the new Supermarine fighter, and was named Merlin shortly afterwards.

The Type 300 was revealed to the public on the occasion of its first flight in March 1936, and by the summer of that year German agents in Britain reported that the RAF had ordered 310 of the new

Messerschmitt's Bf109B was the first fusion of advanced all-metal airframe and European powerplant to enter production. But the big chin radiator still belonged to the era of the Kestrel — with which the Bf109 first flew — and was replaced in the next production version

aircraft as part of its continuing expansion programme. The news arrived at the time when progress with the Bf109 was slow. Evaluation pilots had found that Messerschmitt's new fighter was fast and agile, but difficult to handle at low speed and somewhat unforgiving compared with the contemporary biplane fighters, or with more conservative monoplane types such as the parasol Fw159 or the He112. Developments in Britain, however, caused the faster Bf109 to be viewed more favourably, and in the autumn of 1936 it was judged the best aircraft on offer. Being far easier to produce in quantity than the Supermarine fighter, the Bf109 proceeded rapidly into service. Another factor speeding its development was its more modestly rated, less troublesome engine, coupled to a less advanced cooling system than the Spitfire's.

However, by 1937 Messerschmitt was flying Bf109 prototypes with the much more advanced Daimler-Benz engines and low-drag ducted radiators. Moreover, Daimler-Benz had adopted some new ideas which greatly increased the effectiveness of the new fighter. One was the use of ethylene glycol for cooling, in place of water. Glycol was not only an effective anti-freezing agent, with a low freezing point, but it also had a high boiling point, so that the coolant could be kept at a higher temperature: a smaller volume could cool the same engine, and heat would flow more easily from the radiator to the airstream, so that a smaller radiator was needed. The idea of using glycol was not new –

the US Army had tested it as long ago as 1933, and Rolls-Royce had used it during early flight trials of the PV-12 – but the German industry was the first to use glycol on a large scale. The DB engines had one other unusual feature in common: they were inverted V-12s, a layout chosen because the cowling design permitted a better view for the pilot. Rolls-Royce had originally mocked-up the PV-12 as an inverted engine, for the same reason, but the manufacturers insisted on retaining the Kestrel layout.

The first of the Daimler-Benz engines, the DB600, was built in relatively small numbers. It was followed by the DB601, with two new features: fuel injection directly into the cylinders, and a variable-ratio hydraulic drive to the supercharger. The DB601 set the pattern for the Daimler-Benz engines which powered most wartime German high-speed propeller aircraft. Fuel injection facilitated operation on fuels of low octane, because the gradual introduction of fuel during the compression stroke provided less opportunity for detonation. The variable-speed supercharger was another useful feature. It was automatically controlled to speed up as the aircraft gained height and the outside air pressure fell. Unlike engines such as the Kestrel and early Merlin, the DB601 could be operated at full throttle at all altitudes; engines with single-speed superchargers had a "rated altitude" below which full-throttle operation would lead to excessively high pressures, but this did not concern the Bf109 pilot.

In late 1937 Messerschmitt fitted one of the Bf109

44

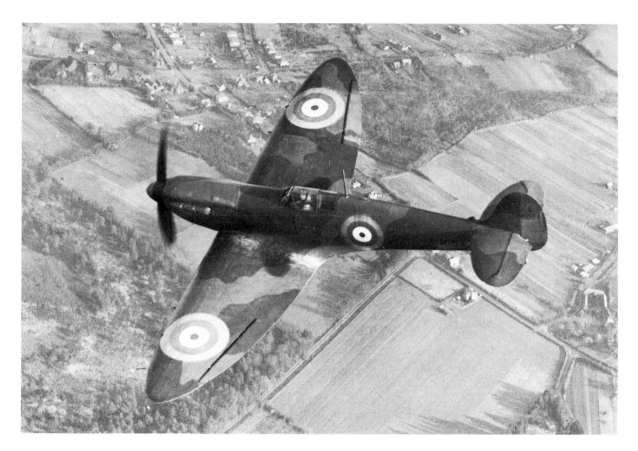

A few years of research, and the discovery of a few key facts, and the radiator has virtually disappeared into the thin, large-area, low-drag wing of the Spitfire. Mitchell's last and most ambitious design was to be one of the fastest fighters in service throughout the war *(Flight International)*

prototypes with a special version of the DB601, which was then starting trials. The DB601R Rekordmotor was modified with internal strengthening to withstand a higher compression ratio and higher speeds in short bursts, and could put out about 50 per cent more power than the standard 1,100hp DB601. In November the Bf109 V13 – which was otherwise closely similar to the DB601-powered production version, the Bf109E – set a new world speed record for landplanes at 379·38mph.

This beat a record which had been set two years earlier by a specially built US racing aircraft, and threw into sharp relief the fact that US service fighters had fallen well behind their European contemporaries. The Army Air Corps had issued a requirement for a new fighter in 1934, as had the Luftwaffe and the RAF, and like the European services the AAC was looking for a modern all-metal monoplane. The Army wanted to start trials in May 1935, but Don Berlin of Curtiss was the only design team leader to meet the deadline and even his Model 75 fighter could not achieve the 300mph speed the Army wanted. The deadline was extended into the following year, making Berlin's revised Curtiss 75B an exact contemporary of the Spitfire. Some 750lb heavier than the Curtiss, with 140hp more power, the Spitfire had four times the firepower and was nearly 65mph faster. What had happened to the US industry?

The best answer was probably a fixation on the radial engine, caused by the fact that there was no

money in making inlines while the services were buying few aircraft and the airlines preferred radials. The dominant engines were two radials– Wright's nine-cylinder, single-row R-1820 and Pratt & Whitney's 14-cylinder, two-row R-1830 – of very similar capacity. They were used on everything from fighters to DC-3s, and were to power every US strategic bomber until 1944. Many remain in service today, which is more than can be said for the Merlin and DB601. But in 1936 their supercharger installations in general use had been developed to airline requirements, and in the absence of pressurisation (then at an early stage of development) their outputs peaked at around 10,000ft. Increasing the rated altitude proved difficult, because the radials tended to be less evenly cooled than a liquid-cooled engine and would not tolerate high pressures and temperatures in the induction system: there had to be some margin of safety to take account of hot-spots in the engine which would otherwise cause detonation. So, although the United States continued to build some of the world's most advanced airframes – the elegant Hughes H-1 distance racer, with its flush-riveted skin and clean design, was an excellent example – and although it led the world in items like propellers, its aircraft lacked the hearty powerhouses of the Old World's fighters.

Meanwhile, as tension increased in Europe, there was still time for a spot of commercial and international bravado. One of the first teams to start were Supermarine and Rolls-Royce, aiming at 410mph and the landplane speed record. The High Speed Spitfire, N17, was fitted with a low-drag windscreen, clipped wingtips, a specially built four-blade fixed-pitch propeller and a special cooling system. The latter avoided the complications of surface cooling by simply allowing coolant to boil off into the atmosphere; a detachable radiator bath was fitted for flight tests. The engine was standard apart from stronger pistons, connecting rods and gudgeon pins, and the fuel was similar to that used in the Schneider race. The modified Sprint Merlin could muster 2,160hp for a record run, and sustained 1,800hp on a 15-hour endurance test.

The Germans, however, put a stop to British ambitions. The first blow came from Ernst Heinkel. Messerschmitt and Heinkel were the unruly barons of the German aircraft industry; neither took much trouble to conceal his unfavourable views on the civil servants and military officers who were nominally in charge of aircraft development, and both were ever ready to cross swords with the Reichsluftfahrtministerium (RLM) on a point of policy, or even to defy direct orders. Heinkel had already shown his independence by developing a completely new high-speed bomber without the RLM's knowledge, let alone their approval.

The bomber, the He119, was highly interesting because despite its size it was among the fastest landplanes of the day, attaining speeds within 10mph or so of the Bf109 V13 speed record. It was radical in general layout and in detail; its powerplant consisted of two DB601s, geared to a single propeller shaft and buried in the centre fuselage. Behind the huge four-blade propeller the fuselage was skinned with perspex, and the two pilots sat one each side of the propeller shaft. Heinkel remained a firm believer in surface cooling, and the He119 was the first of a number of German warplanes to feature such a technique. The system used was more akin to the Goshawk's than to those of the Schneider racers, because the water was actually allowed to boil in the engine and was condensed in the double skin of the wings. A common experiment in elementary physics shows that boiling water to steam takes far more energy than heating it by a similar number of degrees in the liquid state; a corollary is that a steam cooling system needs far less fluid than a liquid system to achieve the same effect, and is accordingly lighter.

The He119 started its trials with a makeshift cooling system on conventional lines, but gradually attained higher speeds as the evaporative system was made to work. The fully developed prototype combined surface condensers with a retractable auxiliary radiator, and attained a top speed of 367mph a few months after the Bf109 set its record. The RLM was unimpressed by the idea of a high-speed bomber without defensive armament, however, and the He119 was relegated to experimental work.

"By the time an aircraft is in service, its replacement should be on the drawing board": it was by this maxim that the procurement specialists of the inter-war period had ensured the availability of new aircraft when they were needed, and in the late 1930s it was still applied. In late 1937 the German industry was asked to consider a replacement for the Bf109; it was hardly surprising that two of the companies beaten by Messerschmitt last time were prominent among the contenders for the new aircraft.

Heinkel's contender drew heavily on experience with the He119, and was the antithesis of the He112 fighter which had lost to the Bf109. The new fighter, for which Heinkel had secured the designation He100 in the RLM sequence, was a small and elegant machine, shrink-wrapped around a DB601 and a surface cooling system. Flown in January 1938, it proved to be very fast indeed; in June one of the prototypes not only beat the Bf109 V13 speed record for landplanes, but did so around a 100km closed circuit rather than a 3km dash, and powered by a stock, unboosted DB601 to boot.

In parallel with the He100 fighter, Heinkel was working on a record-breaking version with a short-span wing, low-drag windscreen and provision for Daimler-Benz's latest Rekordmotor; for the first attempt, this was an 1,800hp powerplant, but the He100 crashed. Another record-breaker was prepared, and was fitted with the new DB601ARJ, running on a lethal concoction of ether and methanol and developing 2,770hp for a period of mere minutes, after which temperatures throughout the aircraft would begin to run out of control. On March 30, 1939, the He100 V8 set a new absolute speed record of 463·92mph, the first to surpass the four-year-old MC.72 record.

But it was Messerschmitt who had the last word. Under the guise of "high-speed research", the professor created the smallest possible aircraft that could contain the DB601ARJ and sufficient fuel for a record run. On April 26, 1939, the resulting Me209 narrowly beat the He100 record, with a 469·22mph run. The pilot, Fritz Wendel, later described his mount as "a vicious little brute" and "a monstrosity", but that did not stop Messerschmitt from spending considerable effort on trying to turn the evil-natured machine into a service fighter. He had no more luck than did Heinkel with the pretty He100; Heinkel's main problem was that he had been too clever for his own good in making use of an existing engine. Urgently needed Bf109s had priority for DB601 deliveries, and the Luftwaffe's next fighter was to be Kurt Tank's Fw190, which made use of a radial engine otherwise fitted to transports.

The official speed record for piston-engined aircraft set by the Me209 was to stand until 1969, not so much because there were no faster aircraft in the category (as we shall see, there were several) but because there were no faster aircraft at the low altitude demanded by the rules for the official record. Absolute speed at sea level or near it was not

to be important, with a few specialised exceptions, in the coming war: rather, the priority requirement was the power to climb and fight at altitude.

It was a time when many new ideas were tried, some of them successfully. There were heights of advanced engineering and feats of applied brute force. Britain, Germany and the United States all favoured different approaches, and to this day there is no clear agreement on the answer to the question: which was the fastest piston-engined fighter?

There is one clear answer, in terms of miles per hour. However, the issue becomes much more interesting if the definition is first narrowed to include only production aircraft or types in full fighting trim, and then widened to include the subsidiary questions: at what time, and at what altitude? From that point, the story includes a great many aircraft, and a great many different solutions to the problems of designing a fighter aircraft and its engine.

One basic point is often overlooked, leading to a tendency to underrate Germany's engineers: German 87-octane aviation fuel was far inferior to that used by the United States' services throughout the war, and by Britain from 1940 onward. US Army Air Corps engineers at Wright Field had realised the potential power benefits of high-octane fuel in the early 1930s, testing a Pratt & Whitney radial on 93-octane fuel to 150 per cent of its normal rating. By 1935, when results of similar tests showed that 100-octane fuel gave the same sort of benefits com-

pared to standard 87-octane, Wright Field campaigned to persuade the US General Staff that the services should switch to the new fuel. The go-ahead was given in the following year, and the new fuel – consisting of Californian gasoline, pure octane and tetra-ethyl-lead – went into production. By 1944-45 the US Army Air Force and the US Navy between them were using 20 million US gal of 100-octane every day.

Germany, by contrast, started the war with 87-octane, used some better fuel at the mid-point of the conflict and was thereafter forced to use an increasing proportion of 87-octane synthetic fuel derived from coal. Additionally, while the quality of the Luftwaffe's fuel declined in the latter stages of the war, the Allies began to experiment with special fuels of even higher performance.

This reversed the situation at the outbreak of war, at which point the Luftwaffe had both the fastest service fighter – in the shape of the Bf109 – and the He100, which was usefully faster than any prototype flying anywhere else, with a 416mph maximum at 16,400ft. However, the industry began to have an increasing amount of trouble with new engines, and the DB601 did not have an enormous amount of development potential, being large in capacity, and relatively lightly built; its moving parts tended to attain excessive speed if it was merely uprated, and efficiency suffered. Its intended replacement was the DB603, but at 44·5 litres capacity this was a very large engine, and it was quite impossible to substitute it for the DB601.

The intended successor to the Bf109 was Tank's compact and formidable Fw190, one of the four outstanding radial-engined fighters of the war in its initial version, and also the only fighter to prove successful with both radial and inline power. A new feature of the engine installation was that the BMW 801 powerplant was cooled by a 12-bladed fan, geared to turn at just over three times the speed of the propeller. Entering service in September 1941, the Fw190 was the fastest fighter of its day – the first to exceed 400mph in service conditions – and proved a shock to the RAF.

The Fw190 was also among the first aircraft to carry the performance-boosting systems with which the Luftwaffe compensated for the inferior performance of its fuel. The first of these to enter service was MW50. Intended to improve the performance of any engine at heights below its rated altitude – where the engine would have to be throttled back to prevent knocking – MW50 was designed to inject a mixture of methanol and water into the supercharger intake. The liquid evaporated, absorbing heat energy in the process and cooling the air before it was breathed in by the cylinder. (The basic theory had been demonstrated in about 1880; Wright Field had looked at it between 1930 and 1941, but it was not until late 1942 that a concern outside Germany, Pratt & Whitney, started investigating it seriously.) The other main boost system developed for the Luftwaffe was GM1, which complemented MW50: while MW50 was used below rated altitude, GM1 was intended to allow the engine to maintain its output despite the steadily thinning air as the aircraft left the rated altitude of its powerplant behind. GM1 comprised an insulated tank of liquid nitrous oxide, fed to the supercharger by compressed air, which acted as an oxidant and boosted the output of the engine by about 30 per cent at above 30,000ft.

The ultimate in boosting, and one of the Luftwaffe's fastest propeller-driven aircraft, was the final development of the Fw190 series, the Ta152H-1 high-altitude fighter. Behind the partially pressurised cockpit there was a GM1 system, in the port wing-root an MW50 set. With the latter in operation the Ta152H-1 could attain 465mph at 29,530ft; using GM1 the maximum speed went to 472mph at 41,010ft, with an ultimate service ceiling of 48,550ft. But by the time the Ta152H went into production in late 1944 there were a number of faster propeller-driven aircraft in limited service.

The single most successful line of development in wartime fighter aircraft was the boosting of the Rolls-Royce Merlin to an ultimate rating more than twice as high as that of the initial 990hp service versions. From 1938, when the Spitfire began to reach the squadrons, there was probably no point at which Merlin-powered aircraft were not among the three or four fastest service types on either side.

The development of new versions of the Merlin started well before the Spitfire even flew. The first step, in January 1935, was the development of a two-speed supercharger drive, which would allow the engine to perform at peak efficiency over a wider range of altitudes. This was not quite as successful as Rolls-Royce had hoped. The company's own two-speed drive gave trouble, so Rolls-Royce took a manufacturing licence on a Farman design; it was longer than the Rolls-Royce type so that the bulk of the engine was moved forwards. The resulting centre-of-gravity shift prevented the two-speed engine from being installed in an unmodified Spitfire, so Britain's fastest airframe had to wait until 1942 for a two-speed-supercharged Merlin.

The next innovation was rather more successful. After testing and rejecting pure ethylene glycol as a coolant, Rolls-Royce in 1936 developed a sealed and pressurised cooling system, taking advantage of the fact that the boiling point of liquids increases with higher pressure. Pressurised to 18lb/in², the Rolls-Royce system allowed coolant temperatures of up to 255°F, similar to those achieved with pure glycol; the benefits in terms of a smaller radiator were similar, but glycol's propensity for finding and exploiting the smallest leak was avoided. The maximum permissible coolant temperature was also maintained irrespective of altitude. Pressurised cooling was exploited in the Merlin XII, which entered service in the Spitfire II in late 1940.

A related development, in which Rolls-Royce's newly formed experimental flight-test department played a major role, was the redesign of the radiator itself (as opposed to the aerodynamic ducting around it). Until 1937, the standard radiator pattern had been the honeycomb type, with hexagonal air tubes passing through a water-filled casing. Rolls-Royce first persuaded manufacturers to switch to a design using smaller tubes, which allowed the weight of coolant in the system to be reduced by 19 per cent. It was then found that even this radiator was based on unsound thinking. Heat, the designers realised, travels more quickly through water than through air, so for maximum efficiency a radiator needs between five and ten times as much metal exposed to the airstream as is bathed by the coolant. The car industry had been using radiators designed to this formula for years: known as "secondary-surface" radiators, they comprised parallel vertical water tubes passing at right angles through sheet-metal fins. Not only did this require 38 per cent less volume of coolant than even the refined honeycomb radiator, but it was cheaper to make and less liable to silt up.

The first aircraft designed from the outset to incorporate such advances in cooling was W. E. W. "Teddy" Petter's Westland Whirlwind fighter, flown in October 1938. New slimline radiators in aerodynamic ducts were buried entirely within the contours of the Whirlwind's notably thin-section wing, being installed between the utterly clean nacelles of the two Rolls-Royce Peregrine engines and the slim fuselage. Combined with the small jet thrust from the ducted radiators, this endowed the Whirlwind with a zero-drag cooling system, or as close to such a system as made little difference. The Whirlwind suffered a worse fate than it deserved, having been designed to a set of requirements which assumed, erroneously, that no single-engined fighter would be able to carry shell-firing heavy cannon. Despite its advanced construction, using extruded wing spars, and the first area-increasing Fowler flaps on a fighter, the Whirlwind was acquired only in small numbers, almost certainly because it was more expensive and complex than single-engined fighters of similar performance. Two years later, the Whirlwind's engine and cooling installation were adapted and scaled up for the Mosquito, with immense success.

Another factor contributing to the Whirlwind's unremarkable service career was that developments of the Merlin family were absorbing most of Rolls-Royce's energies, to the detriment of the V-12 Peregrine and the X-24 Vulture, with which it shared many components. As in the case of the German use of GM1 and MW50, Merlin development followed two parallel routes: higher manifold pressures below the rated altitude, and the improvement of supercharger efficiency to increase power above that height. However, it was assumed that the Merlin would eventually reach its peak at about 150 per cent of its original service rating, and the design of a direct replacement was started in January 1939. The new engine was 36 per cent larger in capacity than the 27-litre Merlin, but by ingenious juggling of components Rolls-Royce managed to design it so that it would fit any Merlin-powered aircraft. (Forty years later this design philosophy resulted in the current world's air speed record for piston-engined aircraft, which suggests that it was a good idea.) The replacement was the Griffon, and development of the Merlin proceeded so well that the new engine supplanted its ancestor in only two production aircraft.

The first step in uprating the Merlin was the adoption of 100-octane fuel, leading to a steady increase in output from basically similar marks of Merlin (the II and III) by mid-1940. As permissible manifold pressures increased, it became possible to run the same engine to higher maximum outputs, these being attained at progressively lower rated altitudes. Finally, these Merlins were delivering up to 30 per cent more power under optimum conditions than the initial production engine could produce on 87-octane fuel. (There is a school of thought which holds that the effect of the American-produced improved fuel on the outcome of the Battle of Britain has not been fully appreciated.) It was found that sodium-cooled valves were mandatory if 100-octane fuel was to be adequately exploited, while new sparking plugs were also necessary.

Development of improved supercharging techniques took longer, but was as rewarding in the long term. Despite all James Ellor's efforts at Rolls-Royce, the performance of superchargers had remained firmly static since the days of the R and the first successful supercharged Kestrel. There were two vital components of supercharger performance: pressure ratio, or the amount by which the supercharger could compress the air, and efficiency, which determined how much power the blower extracted from the engine in the process. Simply increasing pressure ratio – by increasing the speed of the blower, for example – would very often cause such a drop in efficiency that the engine was actually less powerful than before. By 1939–40 service Merlin engines had blowers achieving a 2·3:1 pressure ratio at 65–70 per cent efficiency, which stood comparison with any of their competitors but were little advance on the R and Kestrel blowers.

There were a number of related problems which obstructed Ellor's research in the mid-1930s. The main limit to performance was "surge", the phenomenon that occurred when pressure at the edge of the centrifugal rotor became so high that the blower refused to swallow more air. As pressure ratios were increased, so did the design margins needed to avoid surge, with a consequent loss of efficiency. Another basic problem, which evaded solution for a long time, was that a great deal of energy was being lost in the 180° bend or "elbow" in the duct from the forward-facing ram inlet to the rear centre inlet of the blower. Because the elbow was built into the supercharger casing, the losses in the elbow were not immediately recognised as distinct from losses in the blower; and because a short, sharply angled duct had been developed for convenience of installation, the losses were so large that they tended to mask any improvements that a new supercharger rotor might show. Another obstacle to supercharger development was that inefficiencies and flow breakdowns became much more destructive at higher pressure ratios.

Once variable-pitch propellers had removed one impediment to high-altitude performance, better supercharging became a necessity, and from 1937 more effort was devoted to systematic basic research in the field, including separate testing of the compressor and the inlet. By mid-1940 this research resulted in an improved supercharger delivering a 3·0:1 pressure ratio with five per cent better efficiency than the earlier design. This was adopted first for the Merlin XX two-speed bomber engine and then in a single-speed version for the uprated Merlin 45. The advantage of the latter engine was that it could be installed without too much difficulty in a little-modified airframe, and it was used on the Spitfire V. Many of the latter were converted for low-altitude operation, using a special small-diameter supercharger rotor.

There was one remaining quantum jump in Merlin development. It stemmed from a March 1940 requirement for an engine to power very-high-altitude bombers and interceptors; it was the opinion of some forecasters at the time that the next stage in air combat could involve routine raids from 40,000ft or higher. Combat at such altitudes would clearly require superchargers of very high pressure ratio, and as at that time it was impossible to attain such ratios from a single supercharger rotor, it was decided to use two supercharging stages in tandem. The idea was not new, and by 1940 had been developed to the production stage by Pratt & Whitney in the United States; P&W had started work on the concept in 1934, for the US Navy, and had developed a two-stage-supercharged version of the R-1830 Twin Wasp for the Grumman F4F-3 Wildcat fighter, deliveries of which started a few months after the British high-altitude requirement was issued. However, the American installation had a two-speed drive for only one of the two stages, limiting its performance range, while the aerodynamic technology of contemporary US centrifugal blowers was behind that of the British.

The new Rolls-Royce engine, the Merlin 60, had a two-speed drive for both stages, promising very high pressure ratios in high gear. The two-stage Twin Wasp was fitted with an "intercooler" between the stages, to cool the incoming air and allow higher pressure ratios without detonation; the Merlin 60 had a similar device, except that it was located downstream of the second-stage compressor and

was described as an "aftercooler". In its initial production form, the Merlin 60 could achieve a 4·9:1 pressure ratio at 70–75 per cent efficiency; later versions gave 7·2:1 boost at 62 per cent, and 8·2:1 was attained on an experimental engine.

A pair of Merlin 60s was fitted experimentally to the first prototype of the radical wooden Mosquito. Designed as a high-speed bomber, the Mosquito design had been made possible by the development of powerful synthetic glues and the resulting appearance of very tough, lightweight plywood, and by the rediscovery of strong casein glues which could be "cured" at room temperature and without massive pressure. The Mosquito was largely built of "sandwich" construction, with two very thin plywood skins bonded to a balsa core, one of the main advantages of this material being that it could be formed into aerodynamically uncompromised shapes. For this reason the Mosquito was among the fastest types of its day even though it carried a crew of two and an internal weapons bay. The fastest of all Mosquitos was the re-engined prototype, which attained 437mph in tests.

In the absence of the expected swarms of high-altitude Luftwaffe bombers, the Merlin 60 series was placed in another role: high-speed developments of the Spitfire, which were needed to prevent the Fw190 from upsetting the balance of tactical air

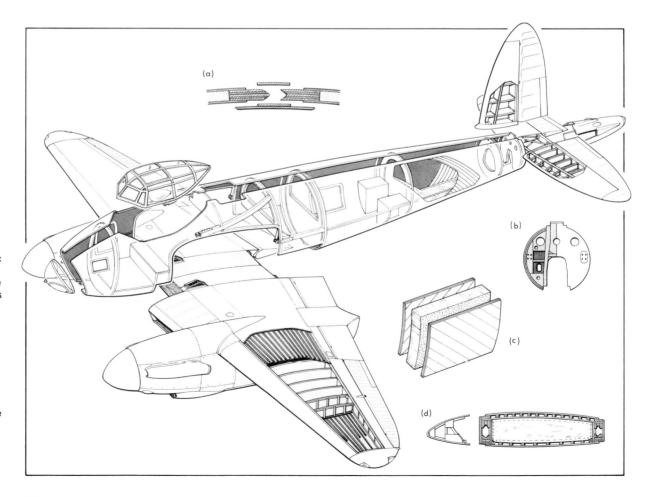

Despite the strong resemblance to a plastic construction kit, this was how the full-size de Havilland Mosquito was put together. The fuselage was built up from ply/balsa/ply sandwich (c) in two half-shells, which were flush-jointed together as in detail (a) around bulkheads of similar construction (b). In the wing skins (d) spanwise stringers of spruce replaced balsa in the sandwich, the main loads being carried by strong spruce spars

superiority over Europe. High power in the thinner air of high altitude gave the Spitfire IX, introduced in mid-1942, the highest top speed of any fighter in service at the time. The Spitfire and the Rolls-Royce V-12s proved to be ideal for each other. Because of the Spitfire's ingeniously designed thin wing (chosen, as was mentioned earlier, to combine large area with low drag) the type was turning out to have a remarkable potential for great speed, particularly at high altitude. The Spitfire IX compared closely in top speed with contemporary American aircraft using engines of 70 per cent greater capacity, such as the Corsair and Thunderbolt. Only when the supercharging techniques developed for the Merlin 60 series were applied to the bigger Griffon did the engine begin to outrun the airframe, but when the first Spitfire with a two-stage-supercharged Griffon entered service in January 1944 it proved to be the fastest fighter in service. By that time, too, Supermarine was well advanced with the design of the Spiteful, the Spitfire's intended replacement, in which the limitations imposed by the airframe were finally removed.

The fastest aircraft to be powered by Rolls-Royce piston engines, however, were not Supermarines. They stemmed instead from a tangled mass of root designs and were brought into the world as a result of a near-conspiracy between US and British officials. Its proto-prototype (the accurate if aesthetically unappealing description) was a fighter designed by Don Berlin for Curtiss, the XP-46. This was basically a refined and much improved descendant of the mediocre P-40, which Curtiss had just started building for the USAAC and for export. While the P-46 was under development, however, the US services needed P-40s urgently, and production of the new fighter was ruled out because it would have caused a gap in deliveries.

At the same time the P-40 was the best fighter for European conditions that was available in the USA, and a British purchasing team was looking for a second source of the indifferent Curtiss. The USAAC engineering people at Wright Field were not even involved in export orders, but Col Oliver Echols saw an opportunity that was too good to miss. In January 1940 he proposed to the British that Curtiss and the Army Air Corps should pass the XP-46 data to another manufacturer, less burdened with vital production work. The British would thus get a better aircraft and the AAC would have some hope of replacing the Curtiss P-40. The chosen company already had a close relationship with the British, as the RAF had ordered advanced trainers of its design: it was one of the West Coast newcomers, not yet grown to the stature of Lockheed, Boeing or Douglas, and it was called North American Aviation.

NAA carried out a complete rework of the Curtiss design. Some features such as its inward-retracting landing gear and its rear-mounted ventral radiator were retained, but the NAA aircraft received a new wing during development. This was the so-called "laminar-flow" wing. Developed by NACA, the laminar-flow wing differed from earlier sections in that the point of maximum thickness was moved further aft, typically to about 40 per cent chord. Its name arose from the fact that the ideal flow across a wing is "laminar", in which the air flows in layers (or laminations) of smoothly varying speed, the fastest stream flowing along the wing surface itself. Even a tiny surface irregularity will cause the flow adjacent to the wing to break up and become turbulent. However, by moving aft the point of maximum camber and air velocity, and hence of minimum pressure and maximum lift, the laminar-flow wing was intended to improve the situation and make the aircraft more efficient. Also, by moving the point of maximum thickness closer to the centre of the wing, the laminar-flow section eased the problem of providing adequate resistance to twisting in the wing. This had proved to be a limitation on the otherwise very fast Spitfire wing, which was of thin section but with a maximum thickness well forward of the aileron hinge line. The ailerons of early Spitfires became very hard to move as air loads increased with high speeds; fitting more powerful ailerons, however, would cause the wing itself to twist before the aircraft would roll. The twist of the wing, in the opposite direction to the deflection of the ailerons, would cause the aircraft to roll powerfully in the opposite direction to the control input. This was one of the first encounters of the time with control reversal, a phenomenon previously confined to some poorly designed high-speed racing monoplanes of the 1920s.

In October 1940, just five months after the NAA team started work on the basis of the XP-46 data, the new NA-73X made its first flight. The aerodynamics of the cooling system needed some refinement, but it was still clear that the new fighter was an immense improvement on the P-40. It was ordered for the Army Air Force (as the Air Corps was renamed in 1941) as the P-51 or, when funds assigned to fighters ran out, as the A-36 "dive-bomber", fitted with wing-mounted airbrakes which were promptly wired shut as soon as the aircraft reached the customer. To the RAF the new aircraft was the Mustang.

The RAF liked the Mustang immensely, but had serious reservations about what was going on – or rather not going on – under the engine cowling. The Mustang had the same powerplant as the P-40: the Allison V-1710, a liquid-cooled V-12 of fractionally larger capacity than the Merlin. There was nothing inherently wrong with the V-1710, except that it had been under development since 1930, usually on a shoestring, despite the fact that Allison was (and still is) owned by General Motors. It had started out as an airship powerplant for the US Navy, and production was just getting under way in 1934 when the USS *Macon* crashed into the Pacific and the entire rigid airship programme was scrapped. The

Army Air Corps, however, had ordered a version of the V-1710 for experimental purposes, while continuing to believe that its production liquid-cooled engines would be of far higher performance. In late 1936 the V-1710-C6 was type-tested at 1,000hp. The development was then side-tracked by the promise of the turbo-supercharger, so that the Allison engine failed to benefit from the years of hard work and research into efficient mechanical blowing that the Merlin had enjoyed. The Army ordered the mechanically blown C13 version of the V-1710 as an insurance against the failure of turbochargers or more advanced engines, and development continued to run well behind that of the Merlin. Not only was its output modest, but its performance fell off with altitude to a greater extent than that of the single-speed Merlins of 1940. The Mustang, with an unimproved engine, reached the RAF at the same time as the Spitfire IX with the two-speed, two-stage Merlin 60.

This was a tremendous opportunity. The Merlin had been quite unexpectedly developed to the point where the Spitfire airframe could not fully exploit it, particularly in terms of range; fuel capacity was a limiting factor, and could not be improved much further in the existing aircraft. But the new British fighter airframes had been designed before the development of the Merlin was foreseen, and were laid out around a completely new generation of engines. The Mustang, however, was a newer design by four to five years than the Spitfire, was at the beginning of its career, and was designed around a Merlin-sized engine. And if this was not enough, in September 1940, while the NA-73 was still a gleam in the corner of NAA's prototype shop, the British Purchasing Commission had reached an agreement with Packard covering the manufacture of huge numbers of Merlins in the USA.

Once the British authorities were persuaded that the mating of the Mustang with the two-stage Merlin was a good idea (a procedure likened to starting a fire with damp newspaper), development went ahead rapidly. The original idea had been Rolls-Royce's, but the USAAF quickly moved into its own parallel programme.

To begin with, the Mustang was bred for endurance rather than speed; but notably, and unlike nearly every other high-speed fighter of the war, it never suffered serious compressibility problems. The first really fast Mustang – probably among the fastest production piston fighters – was produced towards the end of the war, as a result of a US initiative and the joint development of better fuel.

The British had adopted a new method of testing octane – or Performance Number, according to the system used in design by that time – in 1940–41, using a highly supercharged test engine. This method confirmed service experience that some (but not all) fuels had a higher octane rating in some (but not all) engines when run at full power, with a high proportion of fuel to air (a rich mixture) in the

engine. In early 1942 the British established a new fuel specification which laid down requirements for performance in both lean – long-range cruise – and rich power conditions. The new requirement specified 100PN lean performance, and 125PN rich performance, and was written 100/125. This was raised to 100/130 for general use, and the improving supply of catalytically cracked gasolines – of better quality than those obtained by the traditional methods of distillation – helped to meet the demand for fuel.

The process had not finished, because the British now spiked their 100/130 with 2·5 per cent methyl aniline and a little more tetra-ethyl-lead (about 1·6 per cent in all) and created 100/150 grade fuel. This was potent stuff, on which the Merlin RM17 was tested to 2,200hp, more than twice the engine's original design rating. In service it was used to permit a 15 per cent power boost for the fighters assigned to "Diver" patrols against the Luftwaffe's V-1/Fi103 flying bombs. This paved the way for the ultimate P-51.

Rolls-Royce had not been enthusiastic for complicated devices to improve engine performance. Fuel injection, for example, was not well regarded because it robbed the engine of an important advantage: the evaporation of the fuel in a conventional carburated engine cooled the charge entering the cylinder, increasing knock resistance. Another device bypassed by Derby was turbocharging, on the grounds that most of the performance gains could be obtained far more simply by using ejector exhausts. Likewise, it was not Rolls-Royce but Packard which added water/methanol injection to the Merlin, allowing high boost pressures to be used on a high-compression-ratio version of the engine which had originally been designed for long-range use. The resulting V-1650-9 endowed the P-51H Mustang with a top speed of 487mph at 25,000ft, a startling 50mph improvement on the previous production model, the P-51D. This was largely attributable to the 500hp power boost made possible by water/methanol injection, high-grade fuel and the tremendous ability of the Merlin to absorb such development and still perform reliably.

By the end of the war, therefore, the line of liquid-cooled V-12 development in Europe had taken service aircraft to nearly 500mph. The solution favoured from the start by the US Army Air Corps eventually achieved almost the same results. It was apparent, in comparing the V-1710 with the Merlin, that the USA had not put as much effort into the design of supercharger compressors as had the UK. There was a simple reason for this. The British compressors were driven directly from the engine, so that they had to maintain good efficiency along with increased pressure ratio, or the exercise was pointless; in the USA compressor efficiency was accorded lower priority, because the compressor was usually driven by the otherwise wasted energy in the exhaust stream. Apart from some losses

caused by back pressure in the exhaust system, the energy extracted by the turbine and fed to the compressor was free, so its efficient use was not vital. This was the philosophy of the turbo-supercharger – now generally abbreviated to turbocharger – the development and employment of which in aviation have been 99 per cent American.

Turbochargers have other interesting attributes, apart from placing a lower premium on compressor design and allowing people to write TURBO in chunky black letters on the boot lids of their cars. The rotation speed of the turbocharger is independent of the engine speed, so its output is limited only by the energy available from the exhaust and the mechanical limitations of the turbocharger assembly. Moreover, the power of the turbocharger actually increases as outside air density falls off, because there is then a greater pressure gradient between the exhaust and the outside air to which it escapes through the turbine. Therefore, the user of a turbocharged engine does not have a concept of "rated

altitude" to worry about. The blower will just continue to spin faster and deliver the same pressure as altitude increases until something breaks or (more likely) the air becomes too rarefied to cool the engine. (In the mid-1950s, the US Strategic Air Command used stripped-down versions of its turbocharged B-36 bomber to fly reconnaissance missions over "sensitive" areas, attaining heights of more than 58,000ft with jet boost but standard piston engines.) The reverse side of this high-altitude performance, however, is that the output of the turbocharger is limited at low level, while the mechanically driven blower can be geared to deliver any amount of boost that the engine will stand. Therefore, any turbocharged aircraft is most efficient at high altitudes.

The concept of turbocharging appears to have been originated by Alfred Bueichi of the Swiss Brown-Boveri company in 1906, and was applied to a few French warplanes by the Rateau company after 1916. The US Army Air Service was

American engineers adapted a high-compression-ratio version of the Merlin to operate with water injection, and created a service powerplant with 130 per cent more power than the original engine. Its sole application was the P-51H Mustang, the fastest version of the line

sufficiently interested in the idea to buy some Rateau turbochargers, and to compare them with a domestically produced equivalent. The company selected to produce a US turbocharger was General Electric, which had been manufacturing steam turbines for power generation since 1897 and was familiar with basic requirements in the field. Starting in June 1918, the two systems were tested side by side at a camp established high on Pike's Peak in Colorado, the GE product proving superior.

Production contracts for GE turbochargers were cancelled with the Armistice, but research work continued. The year 1921 saw two notable events: an altitude record of 40,800ft, set by a Le Père biplane with GE turbocharger, and the sinking of the surplus German battleships *Ostfriesland* and *Frankfurt* by Brig Gen William Mitchell's turbocharged Martin bombers, flying "above effective anti-aircraft fire". The attention given to turbocharging waned after that time, engines such as the Wright and Pratt & Whitney radials being produced with mechanical blowers. Meanwhile Billy Mitchell's outspoken advocacy of air power, and his unconcealed contempt for most of his superiors, led successively to his demotion on Presidential orders, his court-martial and his resignation from the service. He became a noted public figure; in 1932 he headed a delegation to the Democratic national convention and cast his vote for Franklin D. Roosevelt. Mitchell's influence may have had something to do with the increased interest in turbo-supercharged engines as powerplants for high-altitude day bombers.

In 1932 the advanced two-seat fighter on which Lockheed had been working at the time of the company's bankruptcy emerged from Consolidated Aircraft, which had taken over the design. This aircraft, the Y1P-25, was powered by a Curtiss V-1570 with a new GE turbocharger, and in late 1932 attained 247mph at 15,000ft. It was ordered into production as the P-30A and the later PB-2A, with an improved F-3 supercharger; in April 1936 the PB-2A aircraft attained 274mph at 25,000ft. In the same year the Army agreed to co-operate with GE, Northrop and TWA to explore the potential of higher pressure ratios, using the Gamma single-engined exploration and research aircraft. A highly successful test programme led to the adoption of a similar boosting system for the new US heavy bombers: the Boeing B-17, then undergoing service tests, and the rather later Consolidated B-24. Although the B-17 and B-24 had different engines – the former used Wright's R-1820, the latter P&W's similarly sized two-row R-1830 – both had GE B-series turbochargers, feeding air via an intercooler to a mechanically driven second-stage blower. This "hybrid" arrangement may have been complex, but it meant that the best of both types of blower could be exploited. In conjunction with 100-octane fuel, the boosting system allowed the engines of the B-24 and B-17 to sustain their full outputs up to 25,000ft. A similar but larger system was installed on the B-29.

Not surprisingly, in view of the great success of turbocharging in bomber applications, the US Army was eager to use similar technology in fighters. Following apparently successful tests of a turbocharged Allison V-1710 in a modified P-36 – the XP-37 – the Army ordered service-test YP-37s from Curtiss, and turbocharged prototypes from Bell and Lockheed, all in the course of 1937. Further tests with the XP-37 revealed that the turbocharger installation was simply too complex to be fitted into a single-seat, single-engined fighter at that time, and both the Bell and Curtiss designs were modified to accept mechanically blown engines. Lockheed, however, stuck to the original concept, and in the process created the most sophisticated and advanced fighter of its generation.

At the beginning of 1937 the AAF requested proposals for an interceptor to defend key targets in the USA, with high speed, heavy armament and a very high rate of climb. The combination of speed and armament dictated a twin-engine design; Hall Hibbard of Lockheed and his assistant, Clarence L. "Kelly" Johnson, looked at a number of conventional and unconventional layouts before settling on a twin-tailboom configuration for the new Lockheed Model 22. They were probably influenced by the Fokker G.1 heavy fighter, the first all-metal twin-boom monoplane, which had created something of a sensation on its début at the Paris Air Show in the previous year. In the case of the Lockheed fighter, the twin tailbooms were packed with engines, landing gear, turbochargers and radiators; the "fuselage" housed only the pilot, the armament and the nose landing gear. The small and slender wing was fitted with large area-increasing Fowler flaps, like those of the contemporary Westland Whirlwind. Nevertheless the landing and take-off speeds were expected to be high, and in common with a number of contemporary American aircraft the Model 22 was fitted with a nosewheel or "tricycle" undercarriage. Generally regarded at the time as a piece of American gimmickry, the nosewheel gear was to become accepted as essential as landing and take-off speeds rose and aviation became permanently wedded to the concrete runway. The relatively fragile nature of the nosewheel then became less important, and its added stability and control at high ground speeds rapidly became more so. The Model 22 wing was loaded to some 46lb/ft²; the Bf109, three years earlier, had been criticised and questioned on the score of a wing loading barely more than half as high. One of Britain's pre-war aerodrome-builders was in the habit of testing grass airfields in his Mercedes roadster; if he could drive across a newly laid and graded field in any direction at 60mph, the field was good enough for aircraft use. The Model 22, like many other combat aircraft (notably the B-24 and other US bombers), would demand rather more from its runways.

There were a number of other innovations in the

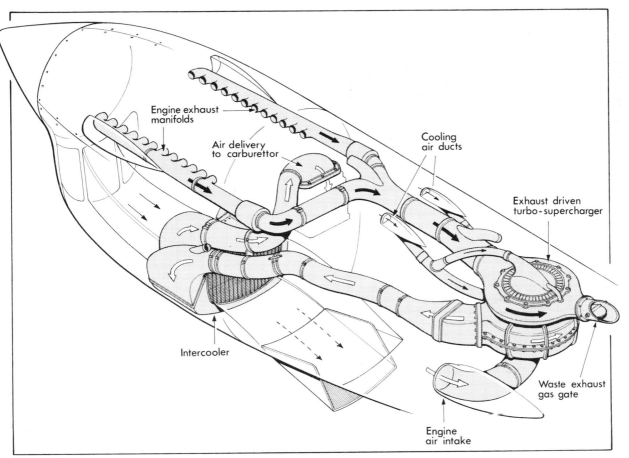

Turbosupercharged engines were efficient but complex, as in this late-model Lockheed P-38 installation. Long, smoothly curved ducts were essential for an efficient gas path, and the Lightning's twin-boom layout was designed to make this possible. The intercooler (sometimes called an aftercooler) was wrapped around the engine-cooling radiator and was almost as bulky, but was needed to cool the highly compressed charge and to allow the engine to run at a high compression ratio without detonation

Model 22. It was a notably clean aircraft, due to the extensive use of flush and butt joints to replace the surface discontinuities caused by overlapping joints. Its Allison engines turned their propellers in opposite directions to improve stability. The General Electric turbochargers were installed in the booms, well aft of the engines, and the compressed air from the blowers was led to the engine via a duct built into the leading edge of each outer wing, acting as an aftercooler to allow higher boost pressures. (This long induction system led to problems, and was eventually replaced with a more conventional aftercooler design with provision for controlling the temperature of the air.) The radiators, neatly ducted, were carried on either side of the booms, leaving the wings clean.

The first Model 22, designated XP-38 by the AAC, was flown in January 1939, and managed a top speed of 413mph – faster than any contemporary combat type except the He100. It was the start of a long, gruelling and sometimes nightmarish development. Weeks after the XP-38 flew, the AAC decided to attempt a transcontinental record flight, but wrecked the aircraft on landing at Long Island. No other aircraft were under construction or even on order, and it was September 1940 before another P-38 flew.

All other problems, however, were overshadowed by the P-38's single most deadly failing: dived from high altitude, the P-38 would suddenly pitch into a steeper dive, and would no longer respond to the elevator. Some P-38s pitched all the way into the bottom half of an outside loop, and became controllable as they climbed and the speed fell away; in some cases pilots who as a last resort had used full elevator nose-up trim in an attempt to recover found that their aircraft began to respond at lower altitudes; other aircraft simply broke up and crashed. The problem had been foreseen by Kelly Johnson in a 1937 memo: the P-38 would reach speeds and altitudes, the designer noted, where the aircraft would be moving at a high fraction of the local speed of sound, or speed of shockwave propagation. It was known at that time that this speed varied with pressure and hence with altitude, and the Mach scale – in which speed is expressed in terms of the local sound velocity, or Mach 1 – was used to some extent to calculate losses in propeller efficiency as the tips reached high speeds. Johnson noted that the XP-38 airframe would probably encounter such effects at around Mach 0·67 – 0·68, equal to about 445mph at 36,000ft.

Knowing why something was probably happening was one thing, but dealing with such a lethal problem was another. Interaction between the wing and the tail was suspected, and a P-38 was modified

Polished and immaculate, the first YP-38 and second Lockheed Lightning passes a top-brass inspection at its roll-out in 1940. Lockheed chief engineer Hall Hibbard, his back to the compact nacelle, talks to US war-production chief W.M. Knudsen (white hat, facing camera) and, in the grey hat, US Army Air Corps commander Gen H. H. "Hap" Arnold *(Smithsonian Institution)*

with cranked booms which raised the tail about 30in; it crashed on one of its first dive tests. Finally, in early 1942, nearly 18 months after the P-38 test programme had got properly under way, tests in NACA's new high-speed wind-tunnel cast light on the problem and pointed the way to a cure. It was found that as the P-38 passed Mach 0·675 the airflow over the wing broke down completely. Normally, the tailplane lay in a stream of air that was deflected downwards by the action of the wing, creating a permanent tailplane download; but this disappeared with the wing lift, causing the nose to tip down. Now, any attempt to pull out of the dive would be self-defeating. If the elevators were to take effect and push the tail down, even slightly, the increased angle of attack on the tailplane would push it up again.

The answer to the problem was to bypass the normal system of pitch control. A pair of electrically powered flaps was installed, one each side of the nacelle, damming the gap between the nacelle and the boom. These simultaneously slowed the P-38 down and created an upthrust to restore normal airflow over the tailplane. They were retrofitted to as many P-38s as possible, and became standard on the P-38J-25. The last-named model also introduced another feature to high-speed aircraft: it had power-boosted ailerons, needed because the heavy P-38 was otherwise slower in the roll than its single-engined adversaries. Powered controls were to make a steady and complete takeover as speeds and air loads increased in later years. In this respect, as in many others, the P-38 was a milestone design in high-speed aircraft, and made a great contribution to solving the problems of the jet age: after the P-38, almost anything must have appeared simple to Lockheed. The final version, the P-38L, could reach high altitudes far faster than the P-51, thanks to a combination of the new Curtiss "paddle-blade" propellers and steady improvements to the type's GE B-series turbochargers. Its powered ailerons helped it to roll as fast as or faster than smaller and supposedly nimbler enemies, and it could extend its Fowler flaps at speeds of up to 250mph if any ill-advised adversary attempted a slow-turning duel.

The P-38 was unique; so was the other turbocharged fighter developed to the orders of the US Army Air Force. This was designed around its powerplant, which dated back to the mid-1930s, when both Wright and Pratt & Whitney could foresee a need for larger engines to power future commercial aircraft. Accepted wisdom held that such engines were too large for fighter applications; the first organisation to question this was the company headed by Major Alexander de Seversky, a Russian emigré who had not let the loss of a leg terminate his career in the Tsar's air force. Shortly after Pratt & Whitney ran its big new XR-2800 eighteen-cylinder radial in mid-1938, the Seversky company offered the Air Corps a massive, long-range, heavily armed fighter designed around an XR-2800 with a large turbocharger in the rear fuselage. It was not accepted. At that time the Army, with a convert's enthusiasm, had embraced the liquid-cooled V-12 (Allison) engine, and had moreover just agreed with Bell and Curtiss that turbocharging for single-engined fighters was impracticable.

Two years later the climate had changed: with the prospect of actual combat, the Army was becoming anxious about the ability of the Curtiss P-40 or Bell P-39 to survive against likely opposition. Alex Kartveli, Seversky's chief designer, had remained head of the design office after the company had been re-organised and been renamed as the Republic Aircraft Company. In mid-1940 the R-2800-powered aircraft was revised and offered again to the Army; this time it was accepted. A pair of ineffectual light fighters under development was scrapped, and the funding from these aircraft – the XP-47 and XP-47A – was quickly switched to the new aircraft, the XP-47B.

The new fighter was built in proportion to the size of its engine. The fuselage was of particularly generous dimensions, to provide a home for three sets of ducts: one duct from the nose to the blower, ducts to feed the compressed air forwards to the engine and exhaust pipes from the engine to the supercharger turbine. Like any turbocharged engine, the combination of R-2800 and the biggest turbocharger that GE or anyone else had built would deliver its peak power at high altitude, so Kartveli did not skimp on wing area. High power and high altitude called for a large propeller, in turn demanding a high and wide landing gear.

Next to any other single-engined fighter of the time, the P-47 Thunderbolt looked like a water-buffalo which some practical joker had entered for the Derby. In the thinner air above 25,000ft, however, the derisive laughter was stemmed, because the chunky, air-cooled Thunderbolt was the fastest fighter of all when it entered service in late 1942. By that time, too, the designers were working on ways to improve even that performance.

It was November 1942 when Republic proposed a refined P-47 to the Army Air Forces. The weight of the airframe would be reduced through structural refinement. Power would be boosted by a bigger supercharger, General Electric's new CH-5. Pratt & Whitney would contribute the R-2800-57 with water injection and a strengthened crankshaft, and a new 13ft-diameter Curtiss propeller would be fitted. The cowling would be redesigned to incorporate a cooling fan as on the Fw190A. This was the XP-47J, and in August 1944 it was officially timed at just over 504mph at 34,450ft, the highest speed recorded for a piston-engined wartime aircraft.

Production plans for the P-47J were turned down at an early stage, because the structural weight-trimming meant changing the bulk of P-47 production tooling for a comparatively small improvement in performance. The fastest production Thunderbolt was the P-47N, using the P-47J powerplant without the cooling fan, which could exceed 470mph on 150PN fuel.

Spitfire, Mustang, Lightning, Thunderbolt: all could claim at one time or another to be the fastest combat aircraft in the world; yet from 1942 onwards there was one family of aircraft which could show them between 20mph and 50mph margin of speed, and which was matched only by the redoubtable P-51H in the last months of the war. Altitude is the qualification. In the fighter-v-fighter or fighter-v-bomber battles above 20,000ft, the types just discussed were supreme, at low altitudes there is another story to tell.

A starting point is the US Army's aeronautical engineering centre at Wright Field, Dayton, where in 1930 a British engineer, S. D. Heron, began to investigate how much power could possibly be extracted from a given engine capacity. The results were remarkable at a time when 600hp or so was the normal service rating of the R-1820 and R-1830. If an engine was run at the high temperatures permitted by pure glycol, and was built with many small cylinders so as to reduce the stroke and speed of the pistons, and was designed to run at high rotational speeds, an output of 1hp/in^3 was feasible. This implied 1,800hp from an engine the size of the two radials just mentioned. Some members of the US Army command became highly enthusiastic about this concept, dubbed the "Hyper" engine, and in the mid-1930s two manufacturers – neither of whom had built a large aero-engine before – were asked to build 12-cylinder prototype engines. They were Lycoming and Continental, names which are not unfamiliar to modern private pilots.

The London suburb of Acton was a far remove from the wide skies of Ohio but gave a home to minds following a similar track. Napier, having missed its opportunity to develop Britain's copy of the Curtiss D-12, was pursuing the development of engines of high speed and high specific output as a means to wrest from Rolls-Royce the dominance of the British market for liquid-cooled engines. By 1930, as was briefly noted in an earlier chapter, a small Napier engine possessing an unusually large number of cylinders had flown in an experimental

Super-Jug: the high-speed development of the Thunderbolt, the 500mph-plus XP-47J, exercises its turbocharged R-2800 on an early ground run. Some of the P-47's apparent ventral bulk is made up of turbocharger ducting passing beneath the wing *(Smithsonian Institution)*

de Havilland monoplane fighter. Like Napier's later engines, it was of H configuration; that is to say, there were two crankshafts geared together, each powered by two opposing banks of four cylinders. Although the two crankshafts could be regarded as a complication, the method was to prove more reliable than a single long shaft carrying all the connecting rods, as in the V-16 or X-24 engine.

This small engine was designed by Maj Frank Halford for Napier and was euphoniously christened Rapier. It was followed by the larger H-24 Dagger; both engines were interesting in that they were air-cooled, yet had little more frontal area than equivalent liquid-cooled engines. By the time the Dagger was developed, however, its 1,000hp output was less than was needed for future fighter types.

By 1935 such future aircraft were envisaged as heavily armed 500mph machines powered by 2,000hp engines, and it was this output which conditioned the design of Napier's new engine. It differed enormously from any other engine of the time. Halford retained the basic layout of the Dagger but laid it flat, and designed it with liquid cooling. Conventional valves were discarded; instead the new engine was to be fitted with sleeve valves, a 1920s invention which had been used on a few luxury cars. The sleeve valve was a steel tube which fitted into

the cylinder, the piston being fitted into the sleeve. The sleeve was driven up and down by a crank and, as it moved, holes in the sleeve uncovered corresponding inlet and exhaust ports in the side of the cylinder.

There were a number of advantages to the sleeve valve. Unlike a conventional valve, it did not generate a local hot-spot to initiate detonation in the cylinder. The combustion chamber could be designed for clean, detonation-free combustion without worrying about the valve location. Both the foregoing meant that the engine could be run at higher compression ratio or boost pressure and would generate more power per unit of capacity. A further advantage was that neither the valve action nor the drive gear need be located on the end of the cylinder, with the result that the frontal area of the engine was small in relation to its capacity.

In November 1935 Napier was given a development contract for this engine, named the Sabre. The planned output of 2,000hp from a 2,240in³ engine was not significantly less than the ultimate aim of the US Army's highly experimental Hyper engine. Considering that it was intended to achieve that output on then standard 87-octane fuel, and the state of the art in supercharging (Rolls-Royce's systematic experiments had not got under way), it can be seen that the basic engine would have to perform in a remarkable manner.

Just over a year after the development contract, Napier was making excellent progress: so good, in fact, that it inspired imitation from no less a proponent of air-cooled engines than Pratt & Whitney. In April 1937 P&W vice-president of engineering George Mead returned from a trip to England and promptly launched development of the X-1800, which despite its designation (denoting "experimental, 1,800hp" rather than its configuration and capacity in the usual American way) was twin brother to the Napier Sabre: a horizontal-H, sleeve-valve, liquid-cooled engine of 2,240in³ capacity. However, Mead resigned in mid-1939 and the X-1800 project lost momentum.

Somewhat similar fates attended the other advanced piston-engines designed for the US Army. By the time the Continental and Lycoming engines were near the stage at which a production decision could be taken, they were too small for forthcoming fighters, while installing them in new versions of existing aircraft would have interrupted vital production for relatively little benefit. The Continental IV-1430 was tested in a modified Lightning, the XP-49, and in a radical experimental fighter developed by a new and little-known St Louis company: the McDonnell XP-67. Lycoming's horizontally opposed engine was paired into a Sabre-sized horizontal-H in an attempt to create an engine powerful enough for a new aircraft, and was tested in a Curtiss naval fighter and the Vultee XP-54.

Chrysler produced an inverted V-16 for the US Army; tested in 1942, it was clearly too late to accomplish anything useful. The weirdest of all was the Wright R-2160, a liquid-cooled engine with a slightly smaller capacity than the Sabre and no fewer than 42 cylinders, arranged in seven radial banks. Known as the Corncob due to its cylindrical shape and plethora of cylinder heads, the R-2160 was scrapped in 1943 because its problematical development was diverting effort from the R-3350 radial, urgently needed for the B-29 Superfortress.

Against this background of failures among radical liquid-cooled engines, Napier's achievement in making the Sabre work appears all the more remarkable. The Sabre has often been characterised as a sick engine, a constant source of difficulties and delays; yet it was the only successful 24-cylinder liquid-cooled engine to see service anywhere, and its development to maturity was faster than that of the much less advanced Wright R-3350. What it always lacked, however, was Rolls-Royce-type supercharger technology to sustain its output to high altitude, and this prevented the aircraft which the Sabre powered from being among the absolute high-speed machines. At low altitude, however, its combination of high power and low frontal area promised tremendous speed, and the new engine was chosen as one of the alternatives for Hawker's new 2,000hp fighter, on which work started in 1938.

In August of that year a plan was hatched to use the Sabre in an attempt to take the world's absolute speed record. Moreover, it would be done on stan-

Britain's Heston J.5 was not only aimed at a 500mph world air speed record, but was intended to do it on standard 100-octane fuel. The Sabre-powered wooden aircraft carried its radiators in the rear fuselage, with its ducts exhausting beneath the tailplane (Smithsonian Institution)

59

In many ways the Napier Sabre was the ultimate high-performance piston engine. The inset detail shows the basic layout, with two horizontally opposed pairs of cylinder banks. The main drawing shows the sleeve-valve system, in which the inlet manifolds and ports (1, 2) and exhaust ports (9) are opened and closed by a thin metal sleeve (3) which contains corresponding ports (4). The axial and rotary movement of the sleeve is controlled by a system of cranks and bearings (5 and 7) and driven by worm gears (6) attached to shafts (8) which run the length of the engine between the upper and lower cylinder banks. The piston and connecting rod (10, 11) move normally inside the moving sleeve. One of the main advantages of the sleeve valve is that the cylinder head (12) and plug (13) design can be extremely straightforward, with an ideally shaped combustion chamber and surrounded by simple and effective cooling passages (14)

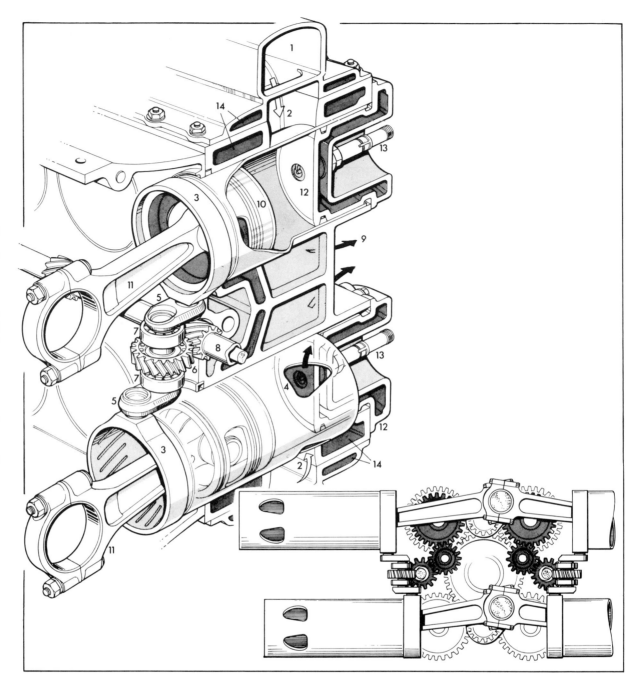

dard gasoline, using a service-type engine, and with an aeroplane built to all civil airworthiness requirements – an immense contrast with the German Rekordmotors, their one-flight lives and their alcohol fuel cocktails. Arthur Hagg, formerly of de Havilland, prepared the basic design for a Sabre-powered record-breaker to be built by Heston Aircraft. The project captured the interest of car manufacturer and philanthropist Lord Nuffield, who financed the aircraft. Construction of two Heston J.5s – later known as the Napier-Heston Racer, although this was never its official name (what race?) – started in 1939. The two German-established speed records of that year did not influence preparations for the record attempt,

because the J.5 was expected to exceed 500mph on a Sabre rated at 2,450hp. As might be expected, it bore in its curvaceous lines a resemblance to Hagg's elegant de Havilland Albatross airliner and, like it, was built of wood. A large ventral scoop fed air to a huge radiator in the form of a shallow forward-facing V, giving maximum radiator area for minimum frontal area.

But the J.5 never had a chance to prove itself. One of the two aircraft was shelved on the outbreak of war. The other took off for the first time in June 1940; seconds later, despite successful ground running, the cooling system failed. (The Sabre's demands on its cooling system were always a problem.) Attempting a dead-stick emergency landing in a new aircraft on Heston's small grass aerodrome, the pilot not surprisingly stalled tens of feet above the field, and the J.5 was wrecked.

The Sabre, meanwhile, was successfully type-tested at 2,200hp for the new Hawker Typhoon. A similar aircraft with the rival X-24 Rolls-Royce Vulture was named Tornado. The Vulture proved totally unreliable, while by late 1941 production Typhoons were being delivered with the reasonably dependable Sabre II. At that time, it was some 700 hp or 40 per cent more powerful than any other engine in production. The Sabre II was rated to deliver its greatest power at about 12,500ft. At this altitude the Sabre could tolerate as much boosting from the supercharger as the Merlin could absorb in thinner air at higher altitude. Cooling and super-charging were the limitations to performance, and when these limits were absent, in bench runs, the Sabre produced phenomenal amounts of power, easily generating 3,750hp and capable of being pushed to even higher outputs.

These ratings were never realised in service, for political and industrial reasons as much as technical causes. By 1941 Rolls-Royce had attained the status of unofficial advisers to the Air Ministry as well as being among its principal suppliers. Beyond all doubt the primary consideration in the minds of the people at Rolls-Royce was producing effective engines as efficiently as possible; but, like all engineers since the Pyramids, they believed that their own products were the best. In the view from Derby, therefore, the national interest and the commercial interests of Rolls-Royce naturally coincided; what really caused bitterness was the fact that the company's opinions carried such immense weight with the Air Ministry.

Soon after the Vulture started to encounter problems, Rolls-Royce had proposed a rationalisation of the engine programme to concentrate on the Merlin and Griffon, both the Vulture and the Sabre being terminated; the move would have cost Rolls-Royce one of its three engine programmes and would have put Napier out of the aviation business completely. The pro-Sabre lobby managed to deflect this move. By May 1942, with the Vulture long dead and buried, Rolls-Royce was still trying to replace the Sabre with the Griffon, which did not match the Sabre's late 1941 in-service rating until 1945. The Air Ministry's controller of research and development noted in that month: "I have the uneasy feeling that this [the Sabre/Typhoon development programme] is suffering through all the propaganda regarding the Griffon Typhoon."

By that time it had been discovered that the effects of near-sonic airspeeds around the Typhoon airframe – which approximated to a scaled-up Hurricane – were a limiting factor on speed, and in late 1941 the Air Ministry ordered a prototype of an extensively revised aircraft, the Typhoon II. This was to be fitted with a thinner-section laminar-flow wing, with Whirlwind-type built-in ducted radiators; new radiators were needed to cool the new high-altitude Sabre IV. The thinner wing provided less fuel capacity, so a fuel tank was added in the fuselage, increasing the fighter's overall length. Most of the resemblance to the Typhoon having vanished, the new fighter was renamed Tempest I before flight tests. In April 1943 the Tempest I

The Hawker Tempest I combined low-drag cooling with a high-altitude Sabre, and would have been Britain's equivalent of the P-47. Development problems meant that the Tempest was used as a low-to-medium-level fighter incorporating the engine installation of the earlier Typhoon (BAe/Hawker Siddeley Aviation)

achieved 458mph at 25,500ft, only a little less than the absolute speed record and a clear 20mph more than any contemporary prototype.

Unfortunately, production of the Sabre was not going according to plan. Napier had been allotted a "shadow factory" at Walton, Liverpool, for the bulk of Sabre production; while the shadow factory system worked well for aircraft and engines which were already developed and established in production, it presented the usual difficulties encountered when preparations are made for production before design changes called for by testing are complete. Neither was the Walton area rich in skilled labour or administrative staff with any experience of such advanced-technology engineering. To avoid further delays, the Air Ministry ordered that production should concentrate on the Sabre II used on the original Typhoon. Together with the Typhoon's chin radiator, this was installed on the main wartime Tempest, the Mark V. In the air-to-ground war at low level, the Tempest V proved invaluable: with a full load of bombs and rockets, it could maintain the same speed as its unladen Spitfire escorts. At just under 5,000ft, the Tempest could reach 416mph in normal service trim, and it was the only propeller-driven fighter which could overhaul the V-1 flying bomb (which cruised at low level) without diving. Towards the end of the war, a few of the even faster Tempest VI were delivered with, at long last, a Sabre rated at higher altitude: with an improved cooling system in the existing chin installation, the Sabre V was rated at 3,055hp, more than the massive turbocharged and water-injected Thunderbolt powerplant.

Changing post-war requirements brought the end of production for the Tempest VI and the Sabre. The main post-war production version of the Tempest was the Mark II, with Sir Roy Fedden's sleeve-valve Bristol Centaurus radial. (The sleeve valve was particularly well matched to the radial engine, because it helped the designer in the eternal struggle to design an efficient and well-cooled cylinder head and drastically cut frontal area.) The Tempest II was not as fast as the Tempest V and VI, which had more power and less drag, but its engine was more efficient in the cruise and it had no vulnerable cooling system, points considered of great importance to the medium/low-level fighter.

The end of Sabre production came much earlier than the appearance of the last examples of its rivals: this was a pity, because by all accounts the Tempest VI could have dominated the new and wild era of air racing which forms a postscript to the wartime story. Before the war, the main prize in US air racing was the Thompson Trophy, awarded by an aviation parts supplier (now known as TRW) for an outright speed race. During the 1930s the speeds attained by the Thompson racers had been slow compared to the later Schneider Trophy speeds, and were little higher than those attained by the latest European service types. The post-war races,

however, were different, being dominated by modified war-surplus fighters. Naturally, the race was run at low level, and in the absence of Sabre-Tempests the fastest type available was an aircraft which the US Navy had designed to chase and destroy Japanese suicide bombers in their final power-dives.

This aircraft, the Goodyear F2G, was something of a monster. It was developed from the fastest wartime carrier fighter, the Chance Vought F4U

Lurking under the cowling of this Goodyear F2G-1D (later designated F2G-2) is the monster 70-litre, 28-cylinder engine which Pratt & Whitney designed for the B-36 intercontinental bomber. The US Navy mated it to the Chance Vought Corsair to produce a kamikaze-catching low-level interceptor (Smithsonian Institution)

Corsair; the work was carried out by the aviation division of Goodyear, better known for its airships, which had been established as a wartime second source for the Corsair. The F2G mated the Corsair airframe with the four-row 28-cylinder R-4360 which Pratt & Whitney had developed for the giant B-36 bomber; the engine was very nearly twice as big as the Griffon or the Sabre. Tuned for racing, the R-4360 yielded an estimated 4,500hp and would pull the Corsair around the course at an average speed of well over 400mph. Although nobody ever tried for the absolute propeller speed record with the F2G, it has been claimed that speeds of 500mph or so were recorded on the straight sections of the course.

Ex-Navy pilot Cook Cleland ran a team of F2Gs and won the race in 1947. For 1948 he added a new weapon in the shape of the sort of "Superfuel" that is always the target of spies in popular fiction. This was triptane, an exotic hydrocarbon first investi-

gated in 1941 and found to yield unbelievable PN ratings; with the addition of a small amount of tetra-ethyl-lead, triptane could be rated at 270PN for high-power operation. There was no engine strong enough to run on pure triptane, but NACA and Allison managed to get 2,800hp out of a bench V-1710 using a blend of triptane with other fuels. The trouble was that production of even a relatively small quantity of triptane – 10 million US gal a day, a few per cent of total fuel use – would have required the entire chlorine production of the United States. A small experimental plant was set up in 1942–45, producing 150gal per day by a method which consumed 2lb of magnesium for each gallon. In 1948 Cleland planned to leave his competitors standing with a 150/200PN-grade methanol triptane fuel from Shell, but the induction system proved unequal to the strain and both Cleland's triptane-fuelled Corsairs retired with their airscoops blown out and their cowlings askew. Cleland won the 1949 Thompson in a less exotically fuelled Corsair, but that was the year that Bill Odom's modified P-51 crashed into a house, killing the occupants as well as Odom and leading to the end of racing at Cleveland.

There is a postscript to the story, and as matters stand now the last word appears to have gone to Rolls-Royce: the postscript takes us to 1964, when Unlimited air racing restarted in the USA at Reno, Nevada. The P-51 Mustangs dominated the scene, for a number of reasons including the fact that they were more readily available than most other types. Also prominent were the Grumman Bearcat – a US Navy interceptor designed as the smallest possible R-2800-powered aircraft – and the Sea Fury, a development of the Hawker Tempest with reduced wing span and weight, and powered by a Bristol Centaurus. It was a modified Bearcat which finally broke the 30-year official record held by the Me209, achieving 483·041mph in August 1969. This record lasted ten years, during which modified Mustangs continued to be the leaders in the annual races. Then came the ultimate modified Mustang: Red Baron Racing's RB-51, powered by a Griffon engine driving a six-blade contra-rotating propeller. In 1979 Steve Hinton took the RB-51 to 499·018mph, which still stands as the official world speed record for piston-engined aircraft; a Sea Fury has been logged at a higher speed, but not officially. The record was set just over 40 years after Rolls-Royce decided to take the trouble to make the Griffon interchangeable with the Merlin, and just about 45 years after the same record was last held by a Rolls-Royce V-12.

The RB-51 was wrecked at Reno in September 1979. More than ever, Merlin-Mustangs dominate the racing scene as other types become more scarce. A pity, perhaps, that the ultimate air race will never be staged: the XP-47J, Tempest VI, F2G-2, Ta 152H and the others. It might, at last, have settled some of the arguments.

Rolls-Royce cunningly designed the Griffon so that it would be only fractionally longer and fatter than the Merlin, but with much more capacity. As a result, it was possible for the Red Baron racing team to put a Griffon into a Mustang to produce the world's fastest piston-engined aircraft (Tom Forrest)

500-600mph

Like an angel's pushing

500–600mph

Like an angel's pushing

AROUND the middle of 1943, some of the most confident Allied pilots were those who flew the photo-reconnaissance versions of the Supermarine Spitfire fighter. They were also some of the least comfortable, because their work involved long flights at high altitude in an unpressurised cockpit designed for 20min fighter missions; but they knew that no aircraft on either side could match their altitude. So it was an unusual experience for one of them to sense a shadow across his cockpit at cruising height. Even more surprising was the source of the shadow: the portly shape of an obsolescent Wellington medium bomber. Downright alarming, however, was another feature of the Wellington; both its propellers were stopped and feathered.

At about the same time, some of the US Army Air Force's height champions – Lockheed P-38s – were training over the California desert when an unidentified aircraft formated on them. This time, the propeller was not stopped; there was no propeller at all, and the UFO was being flown by a cigar-smoking, bowler-hatted gorilla.

Incidents such as these may have caused a decline in bar revenues at the mess for a while, but eventually led to the realisation that, under a shroud of secrecy, some remarkable developments in aviation were taking place. (Anthropoid aircrew were not among them: the apparition which alarmed the P-38 pilots was a quite normal test pilot wearing Hallowe'en headgear.) By mid-1944 these developments were to become a fact of life for service pilots, rather than the stuff of wild rumour.

Radically new ideas were sorely needed, because higher speeds and higher altitudes between them were beginning to mark the end of the development line for the propeller, the simple assembly of spinning aerofoils which had served aviation so well for four decades. Propeller blades presented somewhat thick aerodynamic sections to the air, for structural reasons, and as they rotated they moved through the air rather faster than the aircraft itself. At aircraft speeds roughly half the velocity of sound, or Mach 0·5, propeller efficiency started to decline and an increasing amount of horsepower was required for each further increase in speed. Broader-chord "paddle-bladed" propellers, in which twist was minimised to ease structural problems associated with thin sections, helped at high altitudes but hindered performance lower down; they were mainly applied to high-flyers such as the turbocharged P-38 and P-47.

Some of these difficulties were to be overcome later in high-speed propeller aircraft such as the Tupolev Tu-95 family; more recently, advances in structural materials have permitted the realisation of previously impractical planforms and sections and shown the way to the highly efficient Mach 0·8 prop-fan. At the time, however, it seemed that only by tremendous effort and the extraction of enormous power from ever more complex engines could a conventional aircraft be persuaded to exceed 500mph, even for a short burst.

The ultimate solution to the problem combined a very old idea with a very new one. The old idea was propulsion by reaction, the principle exploited by some cuttlefish which can leap out of the sea by expelling water violently from their bodies. By AD 1500 the principle was being used in man-made devices in the shape of Chinese war-rockets. A man called Wan Hu had the ingenious idea of combining rocket power with aerodynamic technology, creating a flying chariot consisting of two kites, a saddle and 47 rockets. Simultaneous ignition of the latter was a recipe for disaster, and Wan Hu was killed on his first attempt to fly. Rockets – machines propelled by the rapid expulsion of chemically generated gases – continued to be used for military purposes (they included the British Army's Congreve rockets) and by the 1920s a number of pioneers were working on rockets which used mixtures of liquid fuels. However, the first practical applications of reaction power in aviation were the ducted radiator and ejector exhaust stacks developed in Britain in the 1930s.

The other, very new, idea was the gas turbine, a new type of internal combustion engine which was originally conceived not long after the first piston engines were run. Instead of pistons which served in succession as compressors, combustion chambers and means of transforming expansion energy into mechanical power, the gas turbine had a single continuous gas path, so that all phases took place continuously, in different parts of the engine. It consisted in essence of three parts: a blower or compressor, which pressurised the incoming air; a combustion chamber, in which a continuous high-temperature burn was maintained to extract chemical energy from the fuel, and a turbine, which extracted mechanical power from the gas stream and drove the compressor. In the early 1900s Sanford Moss in the USA (later the head of General Electric's research department and a leading figure in turbocharger development) and the French experimenters Lemale and Armengaud succeeded in making gas turbines which would turn unaided, but the efficiencies of the turbine, combustor and compressor were so poor that all the turbine power was needed to drive the compressor – the engine would do no useful work at all. Like all airborne blowers and superchargers of the time, these engines used centrifugal-flow compressors in which air was drawn into the centre of a vaned rotor and flung towards its edge.

The first company to make a success of the gas turbine was Brown-Boveri of Switzerland. From

1905 B-B exploited the principle as a way of using the waste heat from a blast furnace. The furnace, in effect, formed the combustor of a gas turbine, the exhaust gases driving a compressor through a turbine. In 1930–31 B-B applied the same ideas to its Velox boiler for power generation. The early Velox boilers needed some power to help to turn the compressor, but in the late 1930s B-B introduced a new and more efficient compressor design. In effect this reversed the function of the turbine, compressing air between the blades of a fan-shaped rotor. A single such rotor could not achieve the same pressure rise as a centrifugal rotor, so B-B arranged a number of them in tandem on a single shaft. Between each pair of rotors was a "stator", a set of static blades which further compressed and straightened out the airflow from one rotor and fed it to the next. This was the first practical application of the "axial" compressor, and by the end of 1939 B-B had not only improved its boiler design but had delivered 4,000hp gas turbines for an electrical generating powerplant and a locomotive.

The idea of using the gas turbine for aircraft power had been investigated officially as early as 1920 – and resoundingly thrown out. Dr W. J. Stern of the British Air Ministry's South Kensington laboratories prepared a report which was fairly optimistic about the efficiency and output to be expected from a gas turbine driving a propeller, but not about the bulk and weight of such an engine. A 1,000hp turbine engine, Stern calculated, would weigh 6,000lb, one-fifth of that being accounted for by the fuel pump alone.

This damning report did not deter Alan A. Griffith of the Royal Aircraft Establishment from starting work on aircraft gas turbines in 1926. Three years later Griffith obtained a patent on his designs, which were in many respects ahead of their time. As in the later B-B axial compressor, compression took place between the stator blades as well as in the rotor stages, allowing a more compact and efficient engine. However, Griffith's sights were firmly set on propeller-turbine engines, so that his designs had to compete with piston engines on their own terms; and his 1926–29 plans used a somewhat complex and structurally problematical "contraflow" layout in which the turbine and compressor were arranged concentrically and the airflow was reversed through an annular combustor. His efforts were also hampered by the high cost of the facilities needed to test compressors at full scale – a compressor for a 1,000hp turbine would call for a 1,000hp test stand. In 1930 the RAE ordered such facilities from Metropolitan-Vickers (which, like B-B, was in the power-generation business), but in the following year the contract was cancelled to save money. (The RAE/Metrovick relationship survived, and the ultimate offspring of the union was an engine which was still in service at the beginning of 1982.) In 1929 the Air Ministry's Director of Scientific Research described Griffith's ideas as "not worth making

SECRET for military reasons", an official attitude that was to persist for some time.

The 1930s saw a number of individuals and organisations gradually converging on the problems of the aircraft gas turbine. The most attractive feature of the engine was the fact that compression, ignition, power development and exhaust took place continuously and concurrently. Compared with a piston engine, a turbine of the same bulk could breathe more air, burn more fuel and generate much more power. Neither did it have to withstand the sudden explosive force of ignition, so it could be made more lightly. Like a turbocharger – which in effect is a gas turbine using a piston engine as its combustor – a gas turbine is relatively immune to the effects of altitude. On the debit side, its use of energy was far less efficient than the controlled explosive violence of the piston engine, and its fuel consumption per unit of power was much higher. (The comparison still held in early 1982, when advanced gas turbines for airliners were still only just matching the fuel consumption per unit output of later piston engines.) Moreover, the achievement of acceptable efficiency levels meant higher temperatures and pressures. These implied problems of compressor design similar to, but on a much larger scale than, the problems of surge or stall which had faced the designers of superchargers. Then there was the combustor, which had to maintain a constant and intense burn in a roaring airstream, something which had never been done before. Lastly came the turbine, which had to spin at enormous speeds at temperatures of 1,000°F and more, conditions under which most metals would stretch and deform (creep) due to the combination of continuous centrifugal force and heat. And these problems had all to be solved with one eye on weight. Nevertheless, people were seriously looking at the gas turbine as a straight alternative to the piston engine. The breakthrough, however, was the combination of the gas turbine with reaction power. Such an engine would eliminate the weight and complexity of the reduction gear and the propeller, further increasing the power/weight advantage of the turbine. The first patent covering a gas-turbine reaction engine was issued to Guillaume of France in 1921. Near the end of the decade, some far-sighted designers recognised a further advantage – such an engine would be immune to the Mach effects which were beginning to affect most aspects of high-speed design.

After Guillaume of France, the first person to investigate the gas-turbine reaction engine (which we shall, from now on, identify by the name "turbojet", adopted in the USA in the 1940s) in greater depth was Frank (later Sir Frank) Whittle of Britain. Work on a 1928 thesis concerning future developments in aircraft design introduced Whittle to the published work on aircraft gas turbines; in the following year, while serving as an instructor at the RAF's Central Flying School, Whittle conceived the idea of the turbojet. He envisaged from the start a far simpler engine than the Griffith design, with a single centrifugal compressor stage and a single-stage turbine. The Air Ministry's view was that Whittle's assumptions of efficiency and projected performance were highly optimistic, and like Griffith's designs, Whittle's 1932 patent was not considered worth classifying.

The RAF nevertheless recognised Whittle as an outstanding engineer, and by 1934 he was studying mechanical science at Cambridge. Meanwhile, he doggedly pursued his interest in turbojets, and with the help of an ex-service colleague succeeded in raising finance to test the idea. In late 1935 Whittle started the detailed design of an experimental engine, called the Whittle Unit or W.U., and in March 1936 he and his backers formed Power Jets Ltd to test it. The engine was to be built to Whittle's design by the British Thomson-Houston (BTH) company. The decision to proceed directly into development and testing of a prototype turbojet was taken on the grounds of cost, because the company could not afford the expense of the large test rig that would have been needed to drive individual components at representative speeds. The W.U. was intended to develop 1,200lb thrust at 17,750rpm, with a double-sided centrifugal compressor of just

19in diameter. High-pressure air was extracted from the compressor by a single helical duct containing a combustion chamber and fed to a 16·4in-diameter turbine. The design output of this very small engine was similar to that of the biggest contemporary piston engines.

Slowly and erratically, the Air Ministry's attitude towards Power Jets Ltd began to thaw out. To begin with, Whittle was allowed to continue working for Power Jets during his RAF service. By July 1937 Power Jets' funds had run dry, and the Air Ministry supported the company's efforts from that time onwards.

The W.U. made its first test run in March 1937: later Whittle noted dryly that the only speed records were set by the departing observers, as the engine raced out of control on both of its first runs. But the W.U. was later persuaded to run under control (if inefficiently), the first aircraft-type gas turbine engine to do so on conventional fuel. Difficulties lay in the intensity of combustion, which was far greater than any continuous-burning process had ever achieved before. In May 1938 the W.U. was rebuilt with ten individual combustion chambers, or "cans". This alleviated but did not solve the combustion problem. It was found that the burners would either coke up or fill with liquid fuel and

In appearance the early jet engines, such as this Whittle unit, were a far cry from the honed, purposeful-looking contemporary piston engines. Conventional engineers tended to be unimpressed until they realised that the jets were equalling the output of piston engines for a fraction of the weight *(Flight International)*

consequently burn out, but the W.U. nevertheless ran at 16,500rpm for half an hour. About a year later the Air Ministry's Director of Scientific Research, David Pye, visited Power Jets and witnessed a W.U. run. Pye was sufficiently impressed to recommend the development of a flight-qualified engine, the W.1, and an experimental single-engined aircraft to test it, which became the subject of official specification E.28/39 and was ordered from the Gloster company.

By that time a rival jet engine had already flown: the Heinkel HeS 3, designed by Hans von Ohain. As a PhD student at the University of Gottingen, whose aerodynamic research institute (AVA) was to become the centre for German work on compressor aerodynamics, von Ohain was well aware of the Guillaume and Whittle patents, and in late 1933 he designed a simple gas turbine using a centrifugal compressor and its opposite, a radial turbine. The two were mounted back to back on the same rotor, with a combustion chamber on the side. A local car repair firm built the engine under the leadership of Max Hahn; a run was attempted in 1934, but in von Ohain's words it "behaved more like a flame-thrower than an engine". Combustion was highly unstable, much of the fuel burned in the turbine, and the engine could not sustain operation, although it off-loaded its starter motor to a small extent. In the light of these results, von Ohain decided to seek the help of Ernst Heinkel; the latter owned his own company, had no vested interest in the existing type of aircraft engine and was known to be interested in the problems of high-speed flight. In April 1936 both von Ohain and Max Hahn joined Heinkel and started work on an engine designated HeS 1 – the "S" stood for *Sonderentwicklung* or "special development". To facilitate development, von Ohain separated the problems of compressor/turbine design from those of combustion. The HeS 1 was first run on hydrogen, an ideal fuel which presented no mixing difficulties, in April 1937, while in early 1938, an experimental combustor using fuel vaporised before injection was tested successfully. Hahn also invented a reverse-flow combustor that could be installed ahead of the rotor, reducing engine diameter. Another refinement was an axial "inducer" stage ahead of the centrifugal compressor, which was found to ease the load on the main rotor. These features were combined in the course of 1938 in the HeS 3, which in 1939 became both the first turbojet to operate in flight – it was tested in the early part of the year underneath an He118 bomber – and the first to form the powerplant of an aircraft, powering the He178 on its maiden flight on August 27, 1939.

Both Power Jets and Heinkel had by late 1938 succeeded in running turbojets which delivered a useful amount of power, and in this were roughly two years ahead of any other team. However, they were not the only organisations working on turbojets, nor was it generally accepted that the gas-turbine was the sole solution to the problems of high-speed flight. In Germany, Hans Mauch and Helmut Schelp of the RLM had tried in late 1938 to encourage turbojet developments among some of the major engine manufacturers, with mixed results. Junkers and Bramo accepted development contracts (the latter company was acquired by BMW in 1939) for axial-flow engines using technology developed at the AVA in Gottingen. The Junkers Motorenwerke was in an interesting position; in 1935, when the airframe and engine sections of Junkers were separate, the airframe side embarked on the design of an advanced axial-flow turbojet. The RLM contract went to the engine side of the company, and the airframe division's engine was abandoned in May 1939 without running successfully.

As far as the USA was concerned, the Power Jets and Heinkel developments of 1937–39 might have been happening on another planet, despite the fact that energetic research into turbo-superchargers had given the USA a strong lead in high-temperature, high-speed turbines. General Electric's top scientist, Sanford Moss, reported after a visit to the British Thomson-Houston plant that he had seen some large exhaust gas turbines for aircraft use: they were, of course, components of the W.U. turbojet. By 1936, GE's steam-turbine division at Schenectady was working on a large gas turbine for power generation or locomotive use, but the company appeared to share the general impression that gas turbines would be no more efficient and no lighter than conventional engines.

One concept which attracted a great deal of attention at the time was the idea of a piston engine driving a fan, similar to a compressor stage or a cropped propeller, and enclosed in a duct. In some configurations, fuel could be burned in the compressed airstream to generate extra power. Small versions of this system were tested with mediocre results in Germany and the USA, while Italy's Sergio Campini raised support for a prototype. The resulting Caproni-Campini monoplane was heavy, inefficient and slow, and the idea was permitted to perish in the West. A similar system, with an engine-driven ducted fan and reheat, was used in the Soviet Union's MiG I-250 and Sukhoi Su-5 of the immediate post-war years, as a means of increasing the speed of an otherwise conventional fighter.

Other investigators in the late 1930s were following the path pointed out by the unfortunate Wan Hu. Now, however, the liquid-fuelled rocket had been developed to a stage where it could provide power for long enough to make an aircraft practicable; the Opel-Sander experiments with multiple solid rockets earlier in the decade had been little more than a circus act. The foremost figure in the development of rocket aircraft was Helmut Walther of Germany, whose investigations concentrated on the use of highly concentrated hydrogen peroxide (sometimes known as high-test peroxide, or HTP). In high concentrations, hydrogen peroxide can do a

Ernst Heinkel produced some of the most efficient high-speed aircraft to emerge from the Third Reich, though few of them were built in quantity. The He280 was notable as the first jet fighter and the first aircraft to be fitted with an ejection seat, but was passed over in favour of the Me262 *(via Pilot Press)*

great deal more than change the colour of one's hair; under the influence of a catalyst it changes instantly into water and oxygen. This may seem innocuous enough, but the reaction liberates so much energy that the water emerges as superheated steam. The reaction is self-starting, so that any mixing of the fuel and catalyst leads to an instant explosion. Furthermore, the reason that peroxide bleaches hair is that it tends to attack organic material such as rubber, fabric or human tissue.

But the rocket motor promised enormous power for its weight, due to its basic simplicity and the power of its fuel. From 1937 the development of rocket-powered manned aircraft was officially supported in Germany, in a programme that was, and remained, entirely separate from the German Army's rocket-development efforts. In June 1939 the Heinkel He176 made the first flight of an aircraft powered solely by a liquid-fuelled rocket, an 880lb-thrust Walther R I-203. The He176 attained a maximum of only 215mph, but its failure had been foreseen by Dr Lorenz of the DVL (Deutsches Versuchsanstalt für Luftfahrt), which was sponsoring the Walther developments. Lorenz enlisted the help of a young and original designer, Alexander Lippisch, who had built a series of tailless aircraft for the DFS (Deutsche Forschungsanstalt für Segelflug). The tailless design seemed better suited to the compactness of the rocket motor and the need to carry a relatively large amount of fuel. Work on the rocket-powered DFS X and a low-speed

aerodynamic testbed, the DFS 194, advanced during 1938, but the RLM became concerned about the security and other implications of entrusting a potentially important development to the DFS, with its civilian/amateur/academic bent. Lippisch's team was transferred to Augsburg, to form Abteilung L, a special department within the Messerschmitt company.

About the same time Leonid S. Dushkin of the Soviet Union was working on a smaller liquid-fuel rocket motor, and in February 1940 this was tested on a rocket-boosted sailplane. (Perhaps the most interesting thing about the sailplane was the name of its designer; he was Sergei Korolev, the genius who headed Russia's missile and space programme in the 1950s and 1960s.) In both Germany and Russia these first experiments were to lead to the use of rocket power for small, fast-climbing interceptor fighters.

In 1939 the US Navy asked the US National Academy of Sciences to report on the possibility of powering small warships with gas turbines. In January 1941 the NAS submitted its report. Its main conclusion was that the minimum weight for a gas turbine would be around 13lb/hp, so that an engine of equivalent output to a contemporary Merlin would weigh at least three times as much as a fully loaded Spitfire. The weight projection was wrong by a factor of ten to fifteen, even at that early stage, and the NAS report tended to underline the British and German lead.

The RLM's decision to back several turbojet developments proved to be a wise one. By early 1940 three relatively conservative turbojet designs could be selected for accelerated development and possible production. Logically, one of these was a developed version of the von Ohain engine which had flown in August 1939. Designated HeS 8, the new 1,600lb-thrust engine was to be mated with a high-speed airframe designed by Robert Lusser, who left Messerschmitt for Heinkel in 1939 after sketching the outline of a turbojet fighter for the Augsburg company. The new design was a clean, conventional type, and one of the first German aircraft to feature a nosewheel undercarriage. There was one completely new idea on board, indicative of Heinkel's concern for the pilot's safety as speeds increased and air loads made it increasingly difficult to climb out of the cockpit in an emergency: the pilot's seat could be forced out of the aircraft and away from the structure by highly compressed air acting on a piston fitted to the seat. Also, in anticipation of the extreme altitudes of which such an aircraft would be capable, the cockpit of the new aircraft was pressurised, the first application of such a technique to a high-speed aircraft.

The Heinkel team's wide experience with high-speed aircraft helped in the completion of the new He280 by September 1940, before the engines were ready. The aircraft was tested as a glider. In March 1941 the He280 became the second aircraft, and the first operational type, to fly under turbojet power; but engine failures were common and the HeS 8 fell short of its design output. Although the He280 clocked 485mph, this was not startlingly more than the speeds expected from the latest conventional fighters.

Much more impressive speeds were promised from the other engines, of axial layout, and from the aircraft which Messerschmitt was building to use them. Originally known as the P.1065, it was designed around a turbojet which stemmed from the Bramo company, taken over by BMW in 1939. This was the BMW 003, with an axial compressor on the Brown-Boveri/Griffith pattern. The P.1065 was also designed to accept the turbojet on which the Junkers company had started work in late 1938. The basic features of the design had been set by late 1939, when it was designated Jumo 004. It was a deliberately conservative design, drawing on experience with compressor aerodynamics at the AVA at Göttingen, turbine design based on steam turbines produced by AEG, and Junkers' own work on turbochargers and thrust-augmenting exhausts. An unusual feature of the AVA-inspired eight-stage compressor design was that compression took place only in the rotor stages, while the stators served simply to straighten out the airflow. The combustion section was designed to use six separate "can" combustors, simply because the individual combustors were easier to test, using less power, and the process of development was speeded up. The combustor design of the 004 was remarkably free from problems for its day.

The general development philosophy was to proceed as quickly as possible with the testing of an engine which was thermodynamically and aerodynamically representative of the service type, thus sidestepping the absence of test facilities powerful enough to test individual components. The test engine, the 004A, ran in October 1940, and just over a year later accomplished a ten-hour run at 2,200lb thrust, which at that time was far more than any other team was achieving.

The world's fastest aircraft, however, was not powered by the Jumo 004 or by any kind of turbojet. It was driven by one of Helmuth Walther's peroxide rockets, and was setting a pattern which was to have a great influence on the development of high-speed aircraft for decades: the first aircraft to attain any given speed mark were short-endurance, highly temperamental rocket aircraft, and the lessons learned from their brief and perilous flights were to have a great impact on the design of the more conventional machines which followed them. The aircraft which established this trend was the Me163, the new designation of the DFS X project that had been transferred to Messerschmitt in 1938. In 1940 the scepticism aroused by the pathetic hops of the He176 began to evaporate when the DFS 194 was equipped with the same R I-203 engine and attained 342mph at the secret establishment of Peenemünde on the Baltic coast.

Alexander Lippisch had followed a number of investigators in discovering that one way to obtain stability and control without a conventional tail was to rake the wings backwards, and the Me163A – as the DFS X was redesignated – had quite considerable sweepback on its leading edge. The benefits of this were not fully appreciated at the time that the 1,650lb-thrust Walther R II-203b was installed in the Me163A, which had been intended as a pure research aircraft but was now seen as a technology demonstrator for a high-speed interceptor. Between July and September 1941, however, the Me163A began to attain speeds of up to 550mph, and it appeared that the only obstacle to even higher speeds was the limited endurance of the motor, which usually ran out of fuel while the aircraft was still accelerating. On October 2, therefore, an Me163A flown by Heini Dittmar was towed to 13,120ft behind a Bf110 heavy fighter (once again setting a precedent for the assisted launching of research aircraft). Dittmar accelerated until the Me163 pitched violently downwards, then cut the motor and landed safely. Measuring instruments showed that Dittmar had become the first man to exceed 1,000kph. The actual speed was 623·85mph, which at that altitude was equivalent to 84 per cent of the speed of sound. The best wind tunnel then available went to Mach 0.8, and some officials simply refused to believe that the Me163 was so fast. However, this was not the general view, and the

Luftwaffe ordered the rapid development of an operational aircraft based on Me163A experience. Although designated Me163B, this aircraft shared little more than a common concept and configuration with its ancestor.

The production Me163 was modified to overcome what was believed to be the cause of high-speed pitch-down. The wing of the original aircraft was twisted, so that the incidence of the outer wings was less than the incidence at the root. This, combined with sweepback, had been proved essential for a stable tailless aircraft, because at high speed the download on the outermost and rearmost section of the wing simulated the action of a tailplane. However, as the aircraft attained its critical Mach number the outer wing stalled due to the effects of compressibility (at that time the reason why the outer wing stalled first was not understood fully) and, because at such speeds it was producing a marked nose-up moment, the aircraft pitched downwards. Later swept-wing aircraft would do

exactly the opposite. The cure for this was to redesign the wing of the Me163B with a constant degree of sweep across its span. An intriguing and unique feature of the wings of both types was the C-slot, a low-drag fixed slot in the outer leading edge, which accomplished the same function as the more complex automatic slot and did not create measurably more drag.

The other new feature of the Me163B was a completely new engine offering more than twice as much power – 3,750lb thrust for just 220lb weight. While the R II-203 was a "cold" rocket, in which no combustion took place, the new R II-211 used a different catalyst, hydrazine hydrate, carried as a 30 per cent solution in methanol. The methanol base of the mixture burned lustily in the oxygen which the catalyst liberated from the peroxide. The result was an engine which gained much greater power from a given weight of propellant. A small amount of the gas produced by reaction was fed before ignition to turbine-driven pumps which satisfied the enormous

72

appetite of the combustion chamber and kept the rapid reaction stable. Development of the new and immensely advanced engine was slow; the fuel consumption was almost twice what it should have been, and a tremendous amount of work was needed to rectify this. The Me163B airframe was ready in April 1942, but it was not until August 1943 that a powered test flight could be made.

The Me163 was a remarkable technical achievement, but as an operational aircraft it was an abject failure. The new "hot" rocket – designated HWK 109-509A in service – developed more power, but at the expense of an even more lethal mixture of fuels. Any failure on starting the engine could leave the Me163B and its pilot in scattered fragments on the airfield. To save weight for precious fuel the designers had dispensed with a conventional landing gear and substituted a jettisonable two-wheel trolley and skids. If the engine cut soon after take-off –not an uncommon event, because of airlocks and problems with turbulence and cavitation in the fuel pumps – the pilot's chances of survival were slim; because there was no way of dumping fuel in a hurry, the fully loaded Me163B would drop like a stone, and the skid landing gear was not designed for use at high weights. The result of a landing with fuel on board could be an explosion, but this was preferable to a heavy landing which ruptured the peroxide tank and dissolved the pilot alive.

The little Me163B, however, was pleasant enough to fly when these events did not occur, and was by a large margin the fastest operational aircraft of the war, attaining 596mph at 30,000ft. One of the biggest problems was the difficulty of hitting the lumbering USAAF bombers from such a fast-moving platform, using cannon and gunsights intended for a far slower aircraft. This problem was partly solved by the SG 500 Jagdfaust, a battery of vertically fired rockets triggered by a photocell; the significance of this was that it was the first acknowledgement that unaided human reflexes were an inadequate basis for a weapon system at such speeds. The SG 500 was too late to have any material impact on the fate of the Third Reich; so was the rocket fighter which reflected the inadequacies of the Me163B and to some extent rectified them. This was the Me263, with a real undercarriage and a modified engine. The latter was two engines in one: a small "cruising" combustion chamber and nozzle were added to the basic HWK 509, and the fighter could cruise on the power of the small engine alone. This was much more efficient than the boost-glide-boost sequence that the Me163B pilot had to employ and gave the engine fewer opportunities to fail to start. On July 6, 1944, a Messerschmitt Me163B, experimentally fitted with the new engine, reached 702mph, landing safely despite the fact that compressibility had reduced its rudder to a tattered stump.

Two points about rocket aircraft were demonstrated by the Me163: first, that they were very fast, and second, that they were ill-suited to service use. Despite its near-600mph speed, the Me163B's short endurance and lack of a high-speed weapon-aiming device meant that it was of little operational value

On the same day, one month exactly after the invasion of Northern France had begun, another Messerchmitt design also attained the highest speed of its career, setting a mark for conventional aircraft that was not to be equalled for another three years and which was a clear 100mph faster than any Allied aircraft. This was the Me262, originally the P.1065, which had been ordered alongside the He280 in 1940. The design and construction of the Messerschmitt fighter ran even further ahead of its engines than was the case with the Heinkel: by the time that the Me262 V1 neared completion, at the beginning of 1941, the Jumo 004 was delivering barely half its design rating in bench runs.

The technical success of the Me262, which was little short of outstanding given the immense advance in performance which it embodied, was a combination of skill, foresight and an unusual measure of sheer blind luck. Foresight and caution were the main elements of its preliminary design. Robert Lusser, who went on to design the He280, chose to install the engines directly beneath the wings. Although this clearly worsened handling if one engine failed, it avoided the losses caused by the long intake and exhaust ducts of the He178; if the engines turned out to be heavier than predicted, the centre of gravity would not be too badly affected, and the basic aircraft could easily be fitted with different engines. Meanwhile, to reduce frontal area, Lusser designed the aircraft around axial-flow engines, which in the case of the German units then in preliminary design by Bramo and Junkers were notably slim in relation to their length. The nacelles themselves were accordingly long and slim.

Lusser and the rest of the Messerschmitt advanced project office had no means of knowing that they had hit on one of the very few ways of fitting an engine to a wing of normal proportions that would have acceptable characteristics at high subsonic Mach numbers: a long, slender nacelle of roughly even cross-section, installed flush beneath the wing. Later, this "stream-tube" nacelle design was understood, and applied by Boeing to the outer nacelles of the B-47, and two decades further on the same company used it as the key to the problems presented by a new small jetliner, the 737. But the only high-Mach testing of the Me262 was done with a tiny ⅞in model because no larger tunnel was available, and the results were of no practical use.

Good fortune also affected the design of the wing. At a relatively late stage in design it was discovered that the Me262 was turning out to be undesirably tail-heavy, a not surprising error in view of the fact that the Me262 was one of the first high-performance aircraft without the tremendous forward weight concentration of a piston engine. The slow way to fix the problem would have been to move the wing rearwards as a unit. The quick way, adopted by the designers, was the same solution which had been adopted for the Douglas DC-1 transport: the wings were simply swept aft from the

engine mounts to the tips, a not-exactly-streamwise wingtip betraying the nature of the modification. After early trials it was found that the wing behaved better once the leading-edge sweep was made constant over the span. Again, the effects of wing sweep on Mach number were not known, but they were nonetheless there and were a favourable influence on the performance of the Me262.

In other respects the Me262 was *echt Messerschmitt*, of compact design and rugged and simple construction, with high-lift devices more extensive than those of the Bf109 – large slotted flaps and leading-edge automatic slats on the outer wings. Another new feature concerned pitch trimming: while earlier fighter types had invariably used an adjustable tab on the elevator to adjust the balance of the aircraft with changing speed and load conditions, the Me262 featured a variable-incidence tailplane actuated by an electric motor. Pressurisation and ejection seat were absent.

In 1941 the Me262's luck began to run out. The first aircraft was tested with a Jumo 210 piston engine in April, and in November it was flown with a pair of the new and promising BMW 003 turbojets as well. The one attempted flight was terminated by a double flame-out that sent the BMW 003's advanced compressor back to the drawing board. This development left the Jumo 004 as the only powerplant available in reasonable time, but it was not until March 1942 that the engine was cleared for tests under a Bf110; in July 1942 the Me262 V3 was flown with the new engine. By the end of the year, however, the pace of the programme had slowed to a crawl despite the aircraft's potential, for a number of reasons. One was that Luftwaffe procurement by mid-1942 was a shambles, as its entire advanced bomber programme collapsed about its ears due largely to over-reliance on the promises of radical new engines. The replacement for the obsolescent Bf110, Messerschmitt's new Me210, was proving a disastrous flop. The Luftwaffe's procurement chief, Ernst Udet, had shot himself in November 1941, and his responsibilities had passed to Udet's friend and Willy Messerschmitt's old enemy, Erhard Milch. In the circumstances it was hardly likely that Milch would allot high priority to a Messerschmitt design based on a completely new type of engine. Instead, Milch wanted Messerschmitt to concentrate on getting the Me210 right.

Prototype construction of the new jet proceeded slowly. It transpired that the major failing of the Me262 was its tailwheel landing gear. The decision to rectify the centre-of-gravity problem with wing sweep had meant that the mainwheels were well forward, and the Me262 had too little elevator power to raise the tail off the ground. This conspired with an undesirable but inevitable feature of the jet engine: its high-speed propulsive jet was far less efficient at low speeds than a variable-pitch propeller. The result was that the Me262, with its tail down in a high-drag position, could not accelerate

to the point where the elevator would lift the tail. To get the aircraft in a level low-drag attitude so that it could accelerate to flying speed, the pilot had to touch the brakes slightly, jerking the tail upwards in a controlled nose-over. This hazardous procedure was clearly unacceptable for a service aircraft, but Messerschmitt resisted the installation of a nose-wheel gear because it would delay the programme. Eventually the Luftwaffe's insistence on extra payload and fuel forced Messerschmitt's hand: with greater weight and wing loading, the brake-assisted take-off became even more dangerous. The pace of the programme was so slow that it was November 1943 – two and a half years after the first flight on piston power – before an Me262 flew with a retractable nosewheel gear. At that time there were only two Me262s flying, the other being a low-speed test aircraft with a fixed nosewheel.

By that time Junkers had successfully turned the Jumo 004A into the production-model Jumo 004B, in a remarkable development effort spurred largely by Germany's shortage of certain highly important metallic elements: nickel, chromium, molybdenum and cobalt were among the materials which Germany lacked, and which were essential elements of known high-temperature steels. The 004B was not only 220lb lighter than the 004A (a 12 per cent weight saving) but contained only half as much "strategic" material, with a great deal of sheet steel construction. To allow high-temperature components to be made of low-quality material, air was ducted from the compressor and (in the early 004B-2) to cooling slots and pores in the hot rear sections of the engine, including the turbine stator. In the later 004B-4, introduced in late 1944, the turbine rotor blades were made hollow to allow the passage of cooling air, and the engine could thus be made with even less strategic material – late production engines used no nickel at all. The effect of the air passages and pores was to create a film of relatively cool air which clung to the surface of the blade in the same way that water will cling to the curved side of a glass. This formed an insulating blanket between the metal of the blade and the hot gas stream, so that for a given pressure ratio within the engine the metal temperature could be reduced. Air cooling was to be shelved after the war, when strategic metals were readily available, but was to be revived years later as engine pressure and temperature ratios soared beyond the endurance of even the most exotic alloys. Like a piston engine, the gas turbine becomes more efficient, and generates more power for its size and weight, as pressures and temperatures within it are increased; but, unlike a piston engine, it has major components which are continuously soaked in hot gas instead of receiving a cool bath of fresh air

The Junkers Jumo 004 was a notably long and slim engine, a factor which helped in the design of the Me262. Its shape resulted from its conservative, low-risk compressor, which required nearly twice as many stages as other axial compressors

The Gloster Meteor was the only Allied jet to fire its guns in anger. Designed as a fast-climbing, high-altitude interceptor — hence the large wing and tail — it suffered accordingly at low and medium altitudes. Early Meteors were about as fast as a P-51H, with far less range *(via Pilot Press)*

during each working cycle. Of these components, the most critical is the first stage of the turbine. The temperature of the gas stream at the point where it hits this first stage (turbine entry temperature, or TET) must be as high as possible for high efficiency. Advanced versions of the 004 ran at TETs of 1,650°F (1,125 K). This was not far off what had been attained by turbo-supercharger turbines, but was a more difficult target in that the turbojet required a much larger unit under greater mechanical load.

In late 1943 mass production of the Me262 was finally authorised, but at a price. Messerschmitt, seeking to out-manoeuvre Milch, had gone directly to Hitler and persuaded him that the Me262 was an aircraft of enormous versatility. Hitler saw the aircraft as an answer to the de Havilland Mosquito, the *bête noire* of the German leadership. There was, however, one crucial difference between the Mosquito and the Me262. The Mosquito's main armament, whether a cannon pack or bombs, was installed in its weapon bay, close to the centre of gravity. The fearsome battery of 30mm cannon designed into the Me262 was, by contrast, installed in the nose-cone. It could not be removed without upsetting the bal-

ance of the entire aircraft, setting a vital limit on the 262's potential as a bomber. Hitler, despite his keen interest in technical niceties, was not engineer enough to perceive the distinction. Even when the supply of Jumo 004B engines began to pick up, and the production line became organised, the operational use of the aircraft remained in dispute until late 1944.

It was in July 1944 (as noted above, on the same day that an Me163 attained more than 700mph over the Baltic) that a slightly modified Me262, fitted with a special low-drag canopy, attained the remarkable speed of 624mph at Leipheim. Had the aircraft been developed with the same energy and for the same role as the Me163, it *might* have been available in time to mitigate the disastrous loss of air superiority which the Luftwaffe suffered following the introduction of long-range escort fighters. As things were, its impact on hostilities was minimal. Its technical significance, however, was enormous, for it pointed the way to a whole new area of aerodynamic research. In advanced research as much as in operational use, it was the engineers of the Third Reich who led the world in pushing back the barriers of speed.

600-650mph

Arrowheads

600–650mph

Arrowheads

SWEPT wings appeared on the Me262 virtually by accident. Since then, no aerodynamic idea has absorbed so much development effort. So simple in basic principle, swept wings introduce a number of detail complexities into the designer's life; these are a nuisance, but they must be tackled if the designer is to create an aircraft that functions well at much more than half the speed of sound.

Not only did the serious study of sweepback originate in Germany, but so did virtually all the other known solutions to the problem which the swept wing overcomes. That problem is the behaviour of the straight wing at speeds approaching the local velocity of sound. As we saw earlier, high-speed piston-engined fighters were flying fast and high enough to enter the speed and altitude "envelope" in which some of the airflow over the aircraft exceeded Mach 1. The results were violent and often destructive, due to a combination of increased drag, powerful buffeting and – the most deadly factor – uncontrollable changes in trim.

Even the most elaborate wind tunnels in use by the Allies were of only limited usefulness in the investigation of compressibility, because such so-called "subsonic" tunnels were subject to the same physical effects as the aircraft themselves; at high speed the tunnel would choke. However, some useful information emerged from the better tunnels, such as the 10ft by 7ft tunnel opened at Farnborough in 1942. This was one of the best of its day, and could attain a speed of Mach 0·8 at pressures of up to four times atmospheric density – the latter provision reduced the effects of scale. From late 1942 to the summer of 1945 the RAE tunnel was in almost constant use, checking various theories of sonic flow; none of them entirely explained the phenomena encountered. It was found, however, that some supersonic flow might be present over a wing section – indicating that a so-called "critical Mach number" had been reached – before the violent effects classified as "compressibility" were experienced. Most of the undesirable effects could be alleviated by reducing the camber of the wing (the sharpness of the curve over its upper surface) and this in turn implied thinner wing sections.

This was approximately the state of Allied knowledge in the spring of 1945, when advancing ground forces in Germany stumbled across an entirely unknown and brand-new aeronautical research institute at Völkenrode. Allied engineers in the wake of the invaders waxed lyrical about the facilities. "The lavishness in construction and fitting out are most impressive," remarked one envious British researcher. "Unlike most Government institutions it was planned by the people who were going to use it, in this case the scientists, who ...were given a free hand financially." And Völkenrode was only the largest and most modern of German establishments; even the lesser institutions, such as the AVA at Göttingen, stood comparison with the world's best.

It was only in Germany that the behaviour of the sweptback wing had been seriously investigated, although the basic theory had been published in 1935. Adolf Busemann, its originator, was a theoretical aerodynamicist, a species not always trusted by aircraft designers, and his high-speed wing design was based on mathematical principles. If the wing were to be set at an angle to the airflow, Busemann explained, the velocity vector of the airflow would be oblique to the wing. Like any oblique vector, this could be regarded as having two components at right angles to each other: each component would be shorter (slower) than the combined vector, so the critical sonic velocities would not be attained so quickly. So a swept wing can attain the same Mach number as a straight wing of much thinner section (and hence greater weight, all things being equal) before encountering Mach effects. The volume and lift of a swept wing are also distributed along the length of the aircraft, so that trim changes with changing Mach number are likely to be more controllable. For the same reason, a swept wing tends to spread the cross-section of the aircraft along its length, a factor which was later found to be of importance. Also, the swept wing experiences a less sudden rise in drag and change in lift centre with rising Mach number.

The element of spanwise flow which gives the swept wing its desirable qualities creates some problems. Chiefly, the outward-and-rearward flow tends to break down the lift over the outer wings, because air from the roots flows aft and out to the tips, tending to break into the low-pressure zone over the outer wings and causing the tips to stall under extreme conditions. Because the wingtips are well aft, the aircraft will pitch up violently if the tips stall, whether they do it at low speeds or suffer a shock stall, or loss of lift due to compressibility, at high speed. Also, the ailerons become ineffective. Another undesirable swept-wing characteristic is Dutch roll. Because of the geometry of the swept-wing layout, a yaw to the right increases the effective span of the left (advancing) wing, at the same time reducing its sweep relative to the direction of flight, and at the same time causes the reverse effect on the right (retreating) wing. The result is a rolling moment in the same direction as the original yaw, while increased drag on the upward-moving wing dampens and then reverses the original yawing motion. In a large aircraft Dutch roll can manifest itself as a continuous drunken weaving flight that is bad for bombing accuracy and the occupants' viscera. In a small aircraft with lower inertia the phenomenon can be more destructive. Even though these vices were as yet unknown, Busemann's paper was dismissed as Teutonic eccentricity on the international scene, but a second paper of 1937 led to some research at the DVL in Berlin by Dr Waldemar Voigt. In 1940 Voigt joined Messerschmitt as head of the advanced projects office, and the company's own tunnel was used for swept-wing research in 1940–42. The inherent problems of swept wings were clearly recognised, and to begin with there was some doubt whether these could be alleviated enough to produce a flyable aircraft. The palliative chosen by Voigt stemmed from Me262 work: it had been found that the low-speed handling of the Me262 improved when the leading-edge slats (still a Messerschmitt trademark on high-speed aircraft) were extended over the entire span of the wing. Full-span slats were accordingly featured on the 40°-sweep wing of the P.1101, which took shape on the Messerschmitt drawing boards in mid-1942. The wing was also untapered, increasing drag and weight but reducing the risk of tip-stall.

The P.1101 was designed as a high-speed interceptor, powered by a single engine, but development of sufficiently powerful turbojets proved slow. It was not until September 1944 that work on the fighter prototype started, and three months later the requirement for the fighter was dropped. The programme was transferred to Oberammergau, and redirected to pure research (this in Germany, in January 1945!) with a bias towards the low-speed end of the scale. Voigt redesigned the aircraft around a single Jumo 004B, and fixed the wings to a pivot so that the sweep angle could be changed on the ground. The P.1101 was 80 per cent complete by the time the Oberammergau plant was overrun by the Western Allies.

Kurt Tank of Focke-Wulf had also studied swept-wing jet fighters as possible successors to the Me262. Research data on his Ta183 designs may have been acquired by Soviet forces; Tank favoured an untapered wing, but mounted in the shoulder rather than the mid position, and alleviating tip-stall with a device found in a number of German projects – a chordwise strake or "fence" mounted above the wing, which was found to moderate the spanwise airflow at low speeds without compromising the beneficial effects of sweep.

Even bigger surprises awaited the invaders in Junkers' facilities: not one but two swept-wing medium bomber designs. Chief designer Brunolf Baade was well ahead with the design of his EF 150. The 35°-swept wing was shoulder-mounted and drooped towards the tips. The tailplane was set on the very tip of the fin. The landing gear was even odder, with two main units retracting into the fuselage, one in front of the bomb-bay and one behind it, and an outrigger at each wingtip to stop the contraption falling over. And the engines . . . British and American designers gazed in awe at drawings which showed the engines linked to the wing spars by complex cantilever assemblies of steel rods, faired in light alloy, jutting forwards and downwards, so that the rear of the engine lay below the leading edge of the wing.

If Baade's EF 150 gave the impression that the Germans were a nation of mad scientists, Junkers'

In high-speed research the German industry was far ahead of the Allies. At a time when Busemann's original wing-sweep theory was just being dusted off by Allied researchers, the Messerschmitt P.1101, with a swept wing and full-span slats, was nearing completion (*Smithsonian Institution*)

Surely one of the most bizarre aircraft of all time, the Ju287 V1 was doubly significant. It featured the swept-forward wing, uniquely combining low transonic drag, low cruise drag and good characteristics at high angles of attack. It also showed how an unconventional configuration could be tested cheaply at full scale *(Smithsonian Institution)*

other medium bomber – which was much further advanced in development – proved it. The wing of Hans Wocke's Ju287 was swept 25° *forwards*. Design had started in early 1943, and in August 1944 the Ju287 V1 was actually flown. Although the designation indicated that the aircraft was the first prototype of a production design, the Ju287 V1 was a flying low-speed test rig assembled from existing parts – an He177 fuselage, Ju352 mainwheels and a B-24 nosewheel unit unwittingly donated by the US Army Air Force. The wing was all new, a full-scale representation of the production-type wing. The testbed was a large aircraft, and even four Jumo 004Bs would be wholly inadequate to get it airborne. In the absence of larger engines, Wocke installed under each engine a pack containing a Walther HWK 509 rocket and fuel for a few tens of seconds. With two engines flanking the nose, a swept-forward wing and a spatted and fixed landing gear, the Ju287 V1 presented a freakish appearance, even more so than the EF 150. Both, however, were extraordinarily far-sighted designs, and their

bizarre features were all there for good reason. Baade had decided that the EF 150's anhedral wing would help to fight Dutch roll, while the "T-tail" provided additional leverage to combat longitudinal trim changes. The location of the engines was a stroke of genius. Baade had recognised that a turbojet engine did not have the great whirling mass or lateral air resistance of a propeller, and therefore could be mounted less rigidly on the wing. The wing, too, could be made less rigid, and the weight of the engines could be used both to relieve the bending loads on the wing and to add inertia, damping out any rhythmic distortions, or flutter. Most important, however, was that the engines were mounted well forward of the wing, so that there was no transonic interaction between the curvature of the nacelle and the camber of the wing. This avoided the severe "compressibility" buffeting around the nacelle-wing junction which limited many early jets to low Mach numbers. The new-type engine nacelles, of course, could not be used to stow the landing gear; the radical layout of the EF

150 undercarriage allowed the gear to be packed into the slim fuselage, with no worries about providing adequate track.

Wocke had taken a more basic approach to the problems of the swept wing. At high Mach, the beneficial effects of swept wings were apparent whichever way the sweep was applied. The main adverse effect at low speeds – spanwise and aft flow causing local loss of lift – would be reversed, however, so that the roots, rather than the tips, would stall first. Wocke hoped that the result would be a progressive breakdown of lift from the roots outward, and that this could be controlled by inboard slats; full aileron control would be maintained right up to the point of complete stall, at which point the aircraft would simply pitch downwards – the most desirable response. The ailerons could even be drooped to increase lift at low speeds. The main disadvantage, however, was that the wing had to be stiffer and heavier than a sweptback wing: should a swept-back wing flex upwards, air loads naturally tend to force the tips down again, but if a swept-forward wing does the same the air loads will catch underneath it and force it farther up.

Another way of avoiding the worst of the swept wing was originated by R. E. Kosin of the Arado bureau. Arado had produced the straight-wing Ar234, the first jet to be designed as a bomber and reconnaissance aircraft; it had achieved some success, but was little faster than the contemporary piston-engined Do335 Pfeil, due largely to its somewhat conservative design. Kosin developed a new wing that could be applied to the existing Ar234; it was swept 37° at the root, but the sweep-back decreased in steps towards the outer panels, which were swept 25° and carried hinged, drooping leading edges. The object of the leading-edge flaps and the unconventional planform – later known as the "crescent" wing – was to delay tip-stall without prejudice to high-Mach behaviour. The outer wing was made thinner than the root, to compensate for the reduced sweep-back. Kosin's wing was completed but was never fitted to a rocket-boosted Ar234 as had been planned.

There were two other solutions to the problems of swept wings that emerged from late wartime work in Germany. One, related to the ground-adjustable wing of the P.1101, was to provide the aircraft with a means to change its wing sweep-back in flight. This was to be exploited in the immediate post-war years by US experimental aircraft, but never reached the stage of a design study in Germany. One idea that progressed further was conceived by Dr Richard Vogt, the highly creative chief designer of the Blohm and Voss company. Vogt's P.202 design study had an unswept wing, continuous from tip to tip, which could pivot on a central circular bearing so that one wing became swept forwards and one was swept back. Vogt had already defied the convention of symmetry in his BV 141 reconnaissance aircraft and saw no reason why the idea should not work.

Between them, a small group of designers in Germany's advanced project offices had not only managed to find a range of answers to the questions posed by the swept wing; they had produced nearly all the answers that there were. Swept wings were adopted almost immediately by the former Allies in a newly polarised world. Kosin's crescent wing appeared on Britain's Handley Page Victor and (in supersonic form) on the Soviet Union's Tupolev Tu-22. Wocke remained a lone voice in praise of the swept-forward wing, although Tsybin of the Soviet Union tested a manned glider, dropped from a Tu-4, with such a wing, and the first high-speed Ju287 prototype was test-flown in Russia in 1947. Wocke designed the HFB 320 swept-forward-wing business jet of the 1960s. In the late 1970s US designers realised its enormous potential for a high-speed, high-agility aircraft, especially as the availability of new materials had reduced the weight penalty of the swept-forward wing, and at the end of 1981 the US Air Force ordered the Grumman X-29A to test the configuration as a forerunner for a new fighter. Richard Vogt, sadly, did not live to see the successful low-speed testing, in 1981, of a manned aircraft incorporating a centrally pivoted skew-wing, the Nasa/Ames AD-1.

The most immediate problem which confronted the German designers was a straightforward lack of power. In the Jumo 004 Germany's industry had created a jet engine which was reliable enough for service use, could be built in far fewer manhours than piston engines of comparable output and ran on ordinary J2 diesel oil. In the Walther HWK 509 the Germans had the world's most powerful aircraft engine of any kind. In developing lighter, more efficient and more powerful turbojets, however, they enjoyed very little success. By early 1944 the Allies were well on the way to overturning Germany's lead in turbojets, and when the immense booty of German aerodynamic research fell into the invaders' hands there were few obstacles in the way of its translation into operational aircraft.

Much of the credit for the achievements of the Allies can be given to Frank Whittle's tiny Power Jets company. A new fuel injector design had solved many of the company's combustion problems in late 1939, just after the contract for a demonstration engine and a test aircraft had been issued. Early in 1940 results were encouraging enough to justify full approval for a production-type engine and a new high-altitude fighter to use it.

Ordered from Gloster, the company that was building the single-engined E.28/39 test aircraft, the new high-altitude fighter became the Meteor. Its development ran closely parallel to that of the German Me262, but its ultimate performance – except in a highly boosted post-war version – was considerably less inspiring. Conceived as a high-altitude interceptor (at a time when the British Air Ministry was alarmed at the threat of high-level bombing raids), the Meteor was designed using thoroughly conventional high-speed aerodynamics at a time when the phenomena of compressibility were poorly understood. It was less lucky than the Me262 in that its design was based on fat centrifugal engines, and that its designers elected to install them in a mid-position on the wing. This endowed it with a lower limiting Mach number than some of its piston-engined contemporaries, while its large high-altitude wing reduced its speed at lower altitudes.

Another negative factor in the programme was the decision to make another company responsible for production of Power Jets' fighter engine, the W2. (The W1 was the experimental engine for the E.28/39.) The company selected, Rover, had no experience in aviation, but rapidly gained a great deal of confidence in its ability to design jet engines. Relations between Power Jets and Rover's facility at Barnoldswick became strained in the extreme, and by 1942 both companies were trying to cure the engine's problems separately. In November 1942 the W2 flew for the first time, aboard a testbed aircraft converted from a high-altitude Wellington VI bomber. This rate of progress was not considered satisfactory by the Air Ministry, and at the end of the year Rover's Barnoldswick operation was transferred to the control of Rolls-Royce. (From that time onwards, all Rolls-Royce jet engines have been identified by "RB" designations, the "B" standing for Barnoldswick.) Meanwhile two teams had succeeded in overhauling the officially backed Rover organisation, both with direct assistance from Power Jets. One was the de Havilland company, which had been asked in January 1941 to develop a jet fighter. The company decided to build a single-engined fighter around a large 3,000lb-thrust engine of simple layout – the H-1, designed by Frank Halford, the creator of the Napier Sabre. This engine, later renamed the Goblin, also powered the Meteor on its first flight in March 1943.

The other team to make rapid progress was in the United States. In October 1941, a matter of months after an officially sponsored report had concluded that any gas turbine would weigh 13 times its thrust, Power Jets' W1X – the bench-test model of the first flight engine – and complete details of the W2 design were flown to the United States.

By that time, despite the damning report of the National Academy of Sciences, there were a number of US gas-turbine developments under way. Vladimir Pavlecka at Northrop was developing a massive and highly efficient turboprop called the Turbodyne, which Jack Northrop saw as the ultimate powerplant for his huge flying-wing XB-35 bomber. Nathan Price of Lockheed was developing a turbojet engine designated L-1000 for a new all-steel fighter aircraft. Allis-Chalmers was working on a ducted-fan engine, and Westinghouse – also in the power-generation business, and experienced with steam turbines – was looking at an axial turbojet for the US Navy. General Electric's steam-turbine division

at Schenectady had been working on an engine for a PT boat, and had scaled down the design to produce a more easily tested engine. The result turned out to be the right size for an aircraft engine, and formed the basis of the axial TG-100 turboprop. However, it was another GE division – the supercharger unit at Lynn, Massachusetts – which was asked to work on the Power Jets engines. Under cover of developing a colossal turbo-supercharger, GE went to work on the W2 design. In keeping with the letter designations allotted to GE turbochargers, the engine was named Type I.

In June 1942 Whittle travelled to the USA to assist GE in developing the Type I (at that time, Whittle later said, Rover was actually keeping design changes secret from Power Jets) and by September the flight-qualified I-A engine was available. A pair of these was installed in the Bell XP-59A Airacomet, which made the first flight with a W2-type engine in October 1942. The Airacomet was slower even than the Meteor, and its lack of directional stability at high speed ruled it out as an operational fighter, but its high-altitude performance was relatively good, and it proved an excellent testbed for successive GE/Whittle engines.

One of GE's major contributions to engine development was in the area of materials. GE's first engines had forged turbine blades made of a nickel/molybdenum alloy produced by Haynes Stellite Corporation and known as Hastelloy. It was to be replaced from a most unlikely direction. The same company had developed a series of cobalt-based alloys named Stellite, originally as a tool-cutting material. A variant of this had been adopted, as vitallium, by companies that specialised in making false teeth. These were made from a precision original wax model; a plaster cast was built up around the wax, which was then melted out in an oven to leave a high-precision cavity in the shape of the original model. Molten vitallium was then poured into the cavity, and after it had cooled the plaster was chipped away. Lost-wax casting was cheap and consistent, and during the 1940s GE applied it first to the tiny blades of supercharger turbines and finally to the much bigger blades of jet engines, using the Stellite series of metals. Lost-wax casting is now used to produce nearly all turbine blades.

Development of a series of GE engines based on the I-A proceeded rapidly, and in April 1943 the I-16 ran at 1,600lb thrust. Power Jets was at that time achieving similar results with the W2/500, which used a Whittle-designed turbine similar to that of the I-16, while yet a third version of the original W2, the Rolls-Royce W2/B23, was on the point of being approved for flight-testing. The last-named engine was to be the only version of the W2 to be placed in production for an operational type, as the Welland for the Meteor I and II.

The US Army Air Force, however, had decided that the Airacomet or any other aircraft with two I-series (W2-type) engines would be outclassed by future German fighters, and issued a requirement for a much larger powerplant for a single-engined fighter. Frank Halford's Goblin was the most powerful then under development, and Allis-Chalmers was persuaded to drop its own ducted-fan project in favour of building the de Havilland engine. At the same time, GE was asked to produce an even more powerful engine, just over twice as large as the original Type I. The company responded by starting work on two engines, both aimed at 4,000lb thrust. One was the I-40, a centrifugal engine based on a scaled-up I-16 but with straight-through, rather than reversed, combustion chambers for the sake of simplicity. The other was the product of the Steam Turbine Division at Schenectady; the division's TG-100 turboprop had run in May 1943, with some inspiration from the British work, and an axial jet was a logical development. This was the TG-180.

Rolls-Royce, meanwhile, had been following its own line of development, and started by applying its experience in compressor design to a Welland-sized engine using straight-through combustors and a double-sided centrifugal rotor, permitting a greatly increased mass flow for a given engine size. Design of this unit had been started by Power Jets and Rover as the W2/B26, and Rolls-Royce renamed it the RB.26 Derwent. Around the beginning of 1944, however, senior Rolls-Royce officials visiting Lynn witnessed the early trials of the big GE I-40, and realised that both they and the British airframe industry were setting their sights too low. In March serious work started on a new engine closely related to the Derwent, but with double the mass flow and

some new ideas from Power Jets and Rolls-Royce research. The new engine was designated RB.41, later named the Nene, and represented the ultimate wartime development of the Power Jets W2 configuration.

Meanwhile, Britain's Royal Aircraft Establishment and Metropolitan-Vickers had been steadily and slowly working on an axial engine based on Griffith's original design. In June 1941 this was designated the Metrovick F.2, and in late 1943 development started on the revised and uprated F.2/4, this engine running in early 1945 and being designed to yield 4,000lb thrust.

The Western allies, therefore, had four new engines under development in early 1944. All of them weighed about the same as the Jumo 004, then on the point of entering service, but delivered rather more than twice as much power. The problem was that British and American aerodynamic research had not proceeded in step with the tremendous advances in propulsion. The fastest Allied wartime fighters were designed around the early large engines. The first of these was the de Havilland D.H.100 Vampire, flown in September 1943, which attained 530mph on a 2,700lb-thrust Goblin in the following year. The Goblin was also intended to be the initial powerplant for the USAAF's new single-engined fighter; ordered from Lockheed in June 1943, it was designed and built by a hand-picked team headed by Clarence "Kelly" Johnson and flew just 143 days after work started. Just as the XP-80 started flight trials, however, the more powerful GE I-40 started bench runs, proving so successful that

While Germany had the aerodynamics, the Allies had the power, notably a range of 4,000lb-thrust jet engines. General Electric's centrifugal J33 helped the cleaned-up, low-canopied XP-80R to equal the speed set by a similarly modified Me262 three years before *(Smithsonian Institution)*

the USAAF exorcised the Goblin from production plans. Lockheed's fighter had to be redesigned almost completely to take advantage of the new engine; doubtless in practice by this time, Johnson's team designed and built the revised, 25 per cent heavier XP-80A in 139 days, making the first flight in June 1944. The aerodynamics of the P-80 were evolutionary, later versions being fitted with progressively thinner wings for higher Mach numbers. Both the Vampire and the P-80A, however, peaked at about 530-550mph.

Speeds increased slightly in the immediate postwar days. In November 1945 the British installed two highly boosted Derwent Vs in a clipped-wing Meteor and raised the world's air speed record to 606mph. This may seem to indicate that the performance gap between the Meteor and the Me262 was not all that great, but it should be noted that *each* of the Derwent Vs used for the record attempt delivered slightly more power than *two* Jumo 004s. The installation of the Derwent in the Meteor III and subsequent versions of the type improved the critical Mach number, because the cowling was

longer. Shortly afterwards another Meteor attained 616mph. The rules still required a low-level run over a 3km course; an interesting change was the attitude to the weather. While record-breakers in the days of piston engines had waited for cold air to cool their engines, the jets wanted the weather as hot as possible, allowing higher speeds for a given Mach number.

While neither the Nene nor the Metrovick F.2 was to see service on an RAF aircraft, the USAAF enthusiastically adopted both GE engines. Under a new designation system the centrifugal I-40 became the J33 and the axial TG-180 – which had made its first run in April 1944 – became the J35. GE gained little but experience from either engine, because at the war's end the USAAF appointed General Motors' Allison division as sole production source for both the J33 and J35. General Electric's response was to merge the teams which had created the two engines, and set them to build a new axial engine which would fit into any airframe designed around the J35 but develop more power. This was a prudent move, since nearly every American jet then under

The speed record set by the XP-80R was taken back by the Navy/NACA-sponsored Skystreak, a specialised high-subsonic-speed research aircraft powered by GE's axial J35. Note that the tailplane is set at mid-height on the fin to keep it out of the way of shock waves from the wing *(Smithsonian Institution)*

design was being built around the J35 rather than the J33. The smaller cross-section of the J35 was particularly important in multi-engine installations, and proved more attractive than the lighter weight and better handling of the centrifugal engine. The new GE engine was initially designated TG-190, and was ordered into development as the J47 in March 1946.

The stage was set for the fusion of Anglo-American engines with German aerodynamics, and for the appearance of three aircraft which, within the space of a few months, rendered other combat aircraft obsolescent. The new designs were adventurous, because despite all the German research only two swept-wing tailed aircraft had flown by mid-1947: the Bell L-39 piston-engined testbed in the USA and the Soviet Union's Lavochkin La-160. The first of the three service-type swept-wings of 1947, Mikoyan and Gurevich's I-310, crashed on an early flight after demonstrating severe tip-stalling and Dutch roll; while the Russian designers worked on drastic modifications to their prototype, two radical new types flew in the United States. Both owed

a great deal to German design work.

The first to fly, in October 1947, was the North American XP-86. This had started life as a jet development of the Mustang, with a similar wing and new fuselage housing a TG-180 axial engine, which had formed the basis for the Navy's FJ-1 Fury. In 1945 the Air Force approved a mock-up of a similar but refined aircraft, with a thinner wing section and slimmer fuselage, as the XP-86, but no sooner had the prototype been ordered than the first reports of Messerschmitt's swept-wing developments were received. North American quickly redesigned the aircraft to incorporate a P.1101-type swept wing, with slats over most of the span, virtually no taper or twist and 35° of sweep. The ailerons were powered, and the wing was built of double machined skins to avoid any tendency towards aileron reversal. The prototype flew with an Allison/GE J35, but most subsequent models had the more powerful J47. One of the most remarkable things about the new aircraft was the sheer speed of development: the first production P-86A, an entirely workable aircraft, flew in May 1948 and the

First successful swept-wing aircraft to fly was the Sabre. The type's aerodynamics were strongly reminiscent of the P.1101, incorporating the full-span slats first devised for the Me262. Power was provided by GE's new J47. A Sabre was the first production aircraft in history to break the world air speed record

first squadron was formed in February 1949. By this time the newly formed US Air Force had instituted a new designation system, and the aircraft was known in service as the F-86 Sabre. Meanwhile, in May 1948 a P-86A had set a new world's air speed record of 670mph. This broke a record set just nine months before by the Douglas D-558-1 Skystrcak, a purpose-built – but straight-winged – high-speed research aircraft; the P-86A used to set the speed mark was in full service trim.

The Sabre's life-long opponent was to be the aircraft which emerged in late 1947 from six months of intensive redesign into the cold of a Russian winter. The Mikoyan-Gurevich S-01 had prominent chordwise fences to cure tip-stalling, and marked droop on its wings to cure Dutch roll, while to reduce thrust losses in the jetpipe the whole tail had been drastically shortened. The tailplane was carried midway on a highly swept-back fin. The development of the aircraft had been made possible by the British Government, which in 1946 had signed a trade agreement with the Soviet Union; one of the items covered by the agreement was a batch of 20 Rolls-Royce Nenes. The Soviet industry promptly copied the Nene, as efficiently as they had copied a number of interned Boeing B-29 bombers, filling a crucial gap while their own designers caught up on years of missed development: the first home-grown Russian jet engines did not appear in service until the mid-1950s.

Thanks to its small size and the successful uprating of the Nene by Vladimir Klimov's design bureau, the S-01 – which was rapidly placed in production as the MiG-15 – proved a demanding opponent for the Sabre despite the fact that it handled less well at high speed; the Soviet fighter possessed a faster climb rate and a better ceiling.

Both of the new fighters were overshadowed by one aircraft, which was not only nearly as fast as the Sabre or MiG-15 but was one of the world's heaviest aircraft as well. This was the Boeing XB-47, flown in December 1947. Readily recognisable as a descendant of Brunolf Baade's Junkers EF 150, the XB-47 represented an extrapolation of the German designer's ideas, aimed at combining the speed of the jet with the range of the piston-engined aircraft and creating a bomber which would be at least as fast as contemporary fighters.

Boeing was looking at swept wings before German research came to light; in the latter days of the war Bob Jones of NACA had followed the same path as Busemann. At the time Boeing was naturally anxious to ensure that it retained its strong share of the heavy bomber business despite the USAAF's desire to build a fleet of Convair B-36s, but was having difficulty in creating a jet bomber with a sensible range. Sweep-back offered high performance, but posed problems for a multi-engine aircraft because the AAF would not accept the hazards of a multiple installation in the fuselage. Baade's EF 150 design was ideal, and the shoulder-mounted

Mikoyan's MiG-15 was the first of many Soviet swept-wing aircraft to be distinguished by mid-set, drooped (anhedral) wings and prominent overwing fences. The powerplant was an unlicensed copy of the Nene, a Rolls-Royce engine used by almost every air arm in the world except the Royal Air Force *(via Pilot Press)*

The most radical
newcomer of 1947 was
Boeing's extraordinary
XB-47 Stratojet, at once
among the fastest and
the heaviest aircraft of
its day. It was
controversial for years
after its appearance, but
a combination of larger
engines, a system of
inflight refuelling
acceptable to SAC, and
the Korean War led to
orders for more than
2,000 aircraft (Boeing)

anhedral wing, podded engines and bicycle undercarriage were adopted for Boeing's new Model 450. Boeing needed six engines, and originally installed two underwing pairs inboard and one engine on each wingtip, but with the discovery of the "stream-tube" effect it was possible to install the outer engines beneath the wing and to extend the span outboard.

Cautiously, the Army Air Force ordered prototypes of the 600mph bomber. It included dozens of features which had never been tried before. Even six J47s would be inadequate to get the aircraft off the ground, so Boeing installed a battery of solid rockets in the fuselage. All the controls were powered, by a system at unprecedented pressure. Some of the wing skins were made from solid tapered sheet which in places was more than half an inch thick. This was needed to provide adequate stiffness in the wing, which Boeing in its pursuit of aerodynamic efficiency had given the extremely high aspect ratio of 9·43:1. The bicycle undercarriage gave inherently poor braking characteristics, so Boeing devised a system that released and reapplied the brakes as soon as the wheels began to skid. Boeing licensed production and development of the system to Hydro-Aire, and basically similar devices are used on most modern airliners. Early axial-flow jets had poor acceleration, so the B-47 was designed to land

while trailing a parachute. If the pilot wanted to overshoot, he would simply drop the chute and get instant excess power for acceleration. Another, larger parachute was used to slow the aircraft after landing. The B-47's behaviour in Dutch roll was unacceptable for normal flight, so the control circuit was modified to incorporate a gyroscope which constantly sensed the small directional excursions with which the phenomenon begins, and generated automatic correcting moves of the powered rudder. This was the first yaw damper, and also the first mechanical system designed to improve on the natural stability of the aircraft.

Not surprisingly, such an advanced aircraft took longer than the simpler jet fighters to develop and to be accepted for service use. It was the demonstration by Boeing of a practical method of refuelling large aircraft in flight, and the improved performance of the aircraft with J47s, that led to production orders in 1948. Deliveries of ten B-47A Stratojets began in 1950, to be followed by more than 2,000 of the definitive B-47B and B-47E. The US aviation industry had taken a five-year technical lead over its rivals – nowhere was a large jet bomber to be in service before 1955 – and this lead, although it was to be chipped, fractured and sometimes eroded, has never seriously been in question.

88

650-800mph

Jumping the barrier

650–800mph

Jumping the barrier

IT was a British designer, W. F. Hilton of Armstrong Whitworth, who coined the phrase that encapsulated a whole period in high-speed aircraft design and test flying. Hilton described the increasing effects of compressibility, as the aircraft approached the local velocity of sound, to a barrier, either temporary or permanent, which obstructed further progress. An attentive journalist picked up the phrase "sound barrier" and that was that.

"Breaking the sound barrier" is a glib expression, but it was certainly true that in the immediate post-war years, while the jet engine was promising to sustain its power at almost any conceivable speed, it had not been demonstrated that an aerodynamic vehicle could fly safely at more than 80 per cent of the speed of sound. Swept wings, it was true, promised to increase that figure to 90 per cent or so provided that low-speed stability problems could be overcome, and there was a possibility that very thin straight wings might do the same. It was quite possible to build an aircraft strong enough to achieve 100 per cent of the speed of sound, despite the buffeting that affected many high-speed wartime aircraft; it was almost certainly possible to build an aircraft powerful enough, notwithstanding the rapidly increasing drag in the high-subsonic regime; but there was a strong possibility that the behaviour of the air around the aircraft would change so drastically and so quickly that the aircraft might prove uncontrollable.

In fact the state of knowledge on the behaviour of a lifting surface in the transonic or supersonic regime was about the same as the study of lifting surfaces in general had been in the late 19th century. The root causes of the phenomenon of compressibility were understood, but understanding diminished rapidly at the point where compressibility affected a lifting surface. The reason for this was simple: all previous work had concerned the behaviour of bullets and shells, which are simple shapes and develop no lift. As far back as the 18th century Sir Benjamin Robins had detected and measured a force which caused shells to fall short of the range expected from increased charges. In 1892 Charles Vernon Boys managed to capture the flight of a bullet on a high-speed photographic plate: under special illumination it could be seen that the air did not flow smoothly around it but formed V-shaped patterns like the bow and stern waves of a ship.

What was happening, ballisticians concluded, was that any object moving through air creates disturbances all round it, like ripples on a pond. These ripples or pressure waves travel at the same speed as any other waves or pressure pulses in the same fluid: they can be detected by the human ear, so in a sense they are sounds like any others. What the ear senses (as you can hear in the whistle or hiss of a sailplane passing overhead) is an outward ripple caused by the object displacing air as it moves.

But when air encounters an object moving faster than the speed of sound the ripple effect breaks down. The air molecules in front of the aircraft or shell can no longer be displaced fast enough. Instead they are pushed together or compressed, and at the same time the object pushes them forwards at its own speed. Their natural tendency is to move back to their normal volume, which they do. In the process, however, they compress the molecules next to them, which expand and compress the next row. In this way any supersonic object creates an expanding pressure pulse. As the pressure pulse spreads, the object constantly feeds more energy into the centre of the pulse. The pulse expands radially, at right angles to the path of the object: but as it expands the moving object leaves it behind. As the object travels it generates an endless series of pulses, expanding until they encounter the limits of the atmosphere. These pulses, running from the smallest and youngest, just generated at the front of the object, to the largest and oldest, which formed miles back along the path of the object, produce an enormous cone of compressed air, which appears to move at the speed of the object. (In the same way, a wave appears to move on the sea, but in fact the only water movement is up and down.) As this cone "moves over" a pane of glass, the air pressure pushes it one way; as the cone "moves past" the pane the pressure is released and it springs back with a force that can shatter it. To the human ear the wave is readily detectable as an explosive sound.

Artillery shells are designed with sharp noses to generate a classic cone shape. Aircraft are not so simple. Their flight depends on the fact that airstreams reduce in pressure when they are accelerated, as happens over the camber of a wing. As soon as air molecules over the wing or another curved surface reach the speed of sound – something which occurs well before the aircraft does so – the situation changes. Pressure at that point on the airframe increases sharply and lift is destroyed. In the case of most high-speed aerofoils of the 1940s, this phenomenon started well forward of the centre of lift and gravity and resulted in a violent nose-down pitch.

By early 1943 it was understood that thin wing sections helped to delay the worst of compressibility, and that it was most easy to encounter the problem in dives from high altitudes. The thinnest wing on any wartime aircraft was that of the Spitfire (it had been designed to combine low wing loading and long span with low drag) and in 1943 the highest-flying Allied aircraft was probably the new Spitfire PR.XI, a stripped and cleaned-up reconnaissance aircraft powered by a two-stage-supercharged Merlin. The RAE at Farnborough accordingly chose the Spitfire PR.XI as the vehicle for one of the most dangerous series of flight tests in the history of aviation. The RAE's object was to take the Spitfire as high as possible and power-dive the aircraft at successively higher speeds. The trials started in May 1943 and, perhaps amazingly, nobody was killed.

In the course of these trials the RAE developed the first practical Machmeter, the instrument which would thereafter be the main indicator for high-speed flight. Both this and a revised airspeed indicator gave hard data on the speeds recorded in 1943–44. A typical dive would start at 40,000ft or so, with a sharp dive to gain speed. As the speed increased towards Mach 0·74 the pilot would pull the control column back to counter the expected nose-down pitch. This would cut in at about Mach 0·84 and the nose would drop another 10·5° in less than four seconds. From that point onwards the aircraft would accelerate rapidly in the near-50° dive, finally reaching 89 per cent of the speed of sound, or 606mph, at about 28,820ft. At this point the elevators were ineffective, and the pilot actually pushed the stick forwards; as the aircraft descended, its drag in the thicker air began to slow it down, while the speed of sound increased at the lower altitudes. The Mach number accordingly fell off quickly and normal lift was restored. Had the stick been held hard back, or had the aircraft been trimmed nose-up in the effort to pull out of the dive, the Spitfire would have pitched up violently and destructively.

Eventually, one of the diving flights achieved a Mach number of 0·92, at which point the propeller fell off and the pilot accomplished a dead-stick landing. Before that event, however, the RAE had started work on an aircraft intended for research at even higher speeds, preferably without the need for the dangerous and limited diving technique. The aircraft was to be of simple design. The wing was unswept and as thin as structurally possible, while having enough area to provide acceptable high-altitude and landing performance. The tailplane was to revert to the one-piece, all-moving design used on the earliest aircraft, and was to be powered. The fuselage was to be patterned on a rifle bullet, in the absence of any better model of a supersonic shape, while the whole aircraft was to be as small as was consistent with the required thrust and endurance. Specification E.24/43 was written around the new aircraft, and under a shroud of secrecy the detail design and construction of the aircraft were entrusted to the Miles company. The powerplant was to be produced by Frank Whittle's Power Jets organisation.

The M.52, as Miles designated it, emerged as a very small aircraft weighing about 6,000lb in its original form. Its intended powerplant was a derivative of the Power Jets W2/700, mounted inside a duct and driving an "augmentor"; this device consisted of a free turbine in the jet exhaust, with fan blades on its outer rim which drove air through the duct. Downstream of the fan were fuel burners. This form of combustion in the lightly compressed fan

Britain was the first nation to launch a systematic in-flight investigation of transonic phenomena, using a Spitfire PR.XI because of its thin wing and clean design. While the USAAF was spreading nonsensical stories of 700mph dives by P-47s, the RAE's Spitfire was recording genuine Mach 0·92 figures *(Crown Copyright)*

Work on the Spitfire led to the Miles M.52, intended to attain supersonic speed in level flight. The design combined a thin wing with a bullet-shaped body and — in a reversion to early aviation practice — an all-moving slab tail. But the equations of power and endurance kept producing the wrong answers *(Flight International)*

stream (duct-burning, as it is now known) was not very efficient in terms of thrust gained per unit of fuel used, but it promised considerably increased thrust for only a little extra installed weight.

Another use for this type of thrust augmentation – soon known as reheat in Britain, or afterburning in the USA – was to boost the output of an ordinary jet engine. Reheat was tested on Meteors in 1943–44, and showed speed gains of up to 45mph, a 40 per cent improvement in rate of climb and a 65 per cent jump in fuel consumption. The system worked up to 13,000ft – beyond that height it was difficult to obtain stable combustion – and was kept on the shelf in case the Luftwaffe developed a faster version of the Fi103 flying bomb.

The M.52 project rapidly ran into problems. By late 1943 it was becoming clear that it was not possible to get much beyond the Spitfire's dive speed with the W2/700 and augmentor, even supposing that this very advanced and complex power-plant worked reliably and according to predictions. It could be done, perhaps, with a much thinner wing, but then the low-speed behaviour would be so poor that the aircraft would need rocket boost to take off. By July 1944 the RAE regretfully concluded that the aim of going supersonic in level flight was unattainable, especially as the M.52 design had now been enlarged following a design review. Any supersonic flight would be a brief excursion in a dive, and this seemed too little of an advance on previous work to be worth the trouble. It was factors such as this, rather than the publicly stated reason that supersonic flights were considered to be too dangerous for a manned aircraft, which led to the cancellation of the M.52 programme in 1946.

The decision was to prove costly for the British, because the basic design of the M.52 was sound – a scale model was flown to Mach 1·4 in 1948. The loss

of the M.52 experience sent the British industry into a period of uncertainty over the best way of controlling an aircraft into the transonic regime. The traditional approach to the control of a high-speed aircraft had been to tune and balance the control forces ever more precisely as speed increased. There were two main drawbacks to continuing this approach at Mach numbers nearing 1·0. One was the rapid change in longitudinal (pitch) stability as the wing passed through the transonic zone and high-pressure areas formed, this being coupled with violent interactions between the changed airflow patterns and the tailplane – the Lockheed P-38 Lightning had been an early sufferer from the latter problem. The other main difficulty was that increasingly sensitive manual controls became more and more susceptible to flutter and vibration. But the alternative was a complete change of philosophy to embrace a "rigid" system of fully powered controls, with all the questions of complexity and reliability that it would raise.

The British also flirted briefly with tailless aircraft, on the grounds that there could be no problems of interaction between the shock waves from the wing and the tail unit. In May 1946 the de Havilland D.H.108 made its first flight. Based on German research data, the D.H.108 was a tailless aircraft based on a Vampire nacelle married to a 40°-swept wing. What was not appreciated was that the tailless aircraft was weak in pitch damping and control, because the tips of its wings, which provided pitch stability, were too close to its centre of gravity. During practice for an attempt on the world's air speed record in September 1946 the de Havilland D.H.108 crashed into the Thames Estuary, killing its pilot. The flight was unobserved, but the most likely explanation for the crash appears to be an outbreak of uncontrollable pitch oscillation at Mach 0·87–88. The de Havilland D.H.110 two-seat fighter, which had started life as a scaled-up D.H.108, was redesigned with a large tailplane carried on twin booms. The only swept-wing tailless type to see service was the US Navy's Chance Vought F7U Cutlass: the first swept-wing design to be started in the USA, it did not enter service until 1955 and was retired in 1958, to the great relief of those pilots who survived it.

While the M.52 drifted towards cancellation in Britain, the same line of development was being pushed rapidly forward in the United States. Recently published accounts, in fact, have suggested a direct link between the British and American programmes: it was apparently during a meeting between NACA and RAE officials in late 1943 that a Bell engineer proposed that the USA should start work on a transonic manned aircraft. The result was a machine with a marked resemblance to the RAE/Miles design: a thin straight wing, mid-set on a bullet-shaped fuselage with a flush canopy, and a powered irreversible tail, although with a separate elevator.

The future of high-speed research was debated throughout most of 1944, and eventually the US Army Air Force and US Navy went their separate ways. The Navy and NACA concentrated on high-speed jet-powered aircraft, ordering the D-558 Skystreak from Douglas. The USAAF, however, believed that higher performance could be obtained more quickly by using a rocket engine. In November 1944 Bell agreed to develop a rocket-powered aircraft for the AAF, and in the following March three examples were ordered under the designation XS-1.

There were three basic and important differences between the XS-1 and the M.52: it was twice the size, it was designed for pure rocket power and the tailplane was carried at mid-height on the fin, out of the way of anything that the wing wake might try to do to it. Like the M.52, it was designed to take off under its own power and had a conventional tricycle undercarriage.

The engine chosen for the XS-1 was Reaction Motors' XLR-11, burning slightly diluted ethyl alcohol in liquid oxygen – the same mix as the German Army's A-4 rocket – and developing some 6,000lb thrust. However, a serious problem soon emerged, similar to that which had foreshadowed the end for the M.52. One of the most difficult tasks in a liquid rocket engine is to feed the fuel to the combustion chamber fast enough, and yet under control. German engineers, in the Walther and A-4 engines, had managed to produce workable high-speed turbine-driven pumps, but by early 1945 it became clear that such pumps would not be ready for the XS-1. Instead the aircraft was modified to incorporate a high-pressure nitrogen system, comprising twelve small spherical tanks which contained nitrogen at 5,000lb/in^2 pressure. Through a system of pressure-reducing regulators, nitrogen

The Bell XS-1 may have been influenced by the bullet-and-razor shape of the M.52, which was designed a year before it. It was bigger and had a mid-set tail with both variable incidence and conventional elevators (Bell)

was made available at different pressures to drive the landing gear, flaps and tailplane and pressurise the cockpit, at the same time pressurising the fuel tanks with inert gas so that the fuel would be driven steadily through the engine. Thrust variation was achieved by cutting or igniting the four combustion chambers.

The weight of the nitrogen system halved the weight available for fuel; the engine would burn for only 2·5min at full throttle, insufficient for any sort of high-speed testing. The USAAF/Bell team solved their problems by turning to a solution which the British had been forced to reject: air launching. The British had neither the weather nor the runways to facilitate testing a rocket-powered aircraft which had to glide into landing. (Safe operation of such aircraft requires long runways to allow for an efficient, shallow approach path and give plenty of room for long or short landings.) The USAAF had such a facility in the California desert. This establishment was to gain and hold a position in high-speed research somewhat similar to that of Rome in the Catholic Church.

Muroc Air Force Base was named after Muroc Dry Lake, one of a number of such formations in the area. These were large and flat, and although they were occasionally soaked by rain the sun would usually dry them to a concrete hardness. Combined with the clear skies, this made Muroc an ideal location for testing aircraft, particularly those which were expected to have tricky dispositions. It was the availability of Muroc which had enabled GE and Bell to fly the XP-59A Airacomet before its British-designed engines had flown in their home country: under Muroc conditions it was easier to contemplate a dead-stick landing or to operate with less-than-design thrust. In addition to Muroc AFB, the USAAF had plenty of Boeing B-29 Superfortresses which could happily carry the XS-1 to a suitable launching altitude.

The first of the XS-1s was rolled out in December 1945. This aircraft was fitted with an eight per cent thick wing; that is to say, the maximum thickness was eight per cent of the chord. The second aircraft was to be tested with a ten per cent thick wing for comparative purposes. The tail assemblies of the two aircraft were of slightly thinner section than the wings, so that the tail would still be working properly when the wing encountered compressibility.

The thinner-winged aircraft started gliding trials

Vital to the success of the XS-1, and to other US high-speed programmes, was the experience gained in air-launching the aircraft over the California desert. Most of the key speed marks since Yeager's Mach 1 flight have been achieved first by air-launched, rocket-powered US aircraft *(Bell)*

at Pinecastle AFB, Florida, in January 1946, in the hands of Jack Woolams. The aircraft was then grounded for installation of the XLR-11, and the programme moved to Muroc in September. Woolams had been killed in August while practising for the Thompson Trophy race, and his place was taken by another Bell test pilot, Chalmers "Slick" Goodlin, who made the first powered flight in the thicker-winged No 2 aircraft in December 1946.

Under Bell's contract with the Air Force the XS-1 was warranted to be stable up to Mach 0·8. Goodlin made 20 powered flights in 1947, steadily pushing the No 2 closer to the buffet boundary – the thin-wing No 1 aircraft was still not back to flying status, having been damaged in an early test at Muroc. By mid-1947 the USAAF chief of flight testing, Col Al Boyd (who at that time held the world's air speed record in a modified P-80), was becoming impatient with the pace of the Bell/NACA/AAF programme, which appeared to have stopped just before the critical Mach number of the XS-1 was reached. Negotiations were in progress over the six-figure bonus which Goodlin would receive for attempting supersonic flight. Boyd proposed a shorter route: AAF test pilots would take over the No 1 aircraft and attempt to break the sound barrier as soon as possible. This was approved, and Boyd chose Capt Charles "Chuck" Yeager to lead the test programme.

Yeager was a former P-51 pilot who had had a short but eventful war in Europe; arriving in mid-1944, he claimed two victories on his first eight missions, evaded capture after being shot down on the ninth mission and ended the war as a double ace. After the war he had been selected for AAF test duties. His back-up on the AAF XS-1 programme was Lt Bob Hoover, who is likewise not entirely unknown in flying circles today. Both pilots joined the programme for their regular flying pay.

Yeager had three glide flights in which to familiarise himself with the XS-1 operating procedures. On a fully fuelled flight the pilot would stay in the B-29 until the mother ship reached 12,000ft, because at its 13,000lb fuelled weight the XS-1 would stall at 240mph. If it was dropped below 12,000ft the result would be a spin with no time to recover. (There was no ejection seat, and Yeager afterwards noted that the seat-type parachute "was primarily to sit on". The B-29 would climb in widening circles, staying within gliding distance of Rogers Dry Lake, and would then move out to 40 miles from the landing strip at 25,000ft. At this point the top level speed of the laden B-29 was less than the stall speed of the XS-1, so the bomber went into a 15° – 20° dive before releasing the research aircraft. After the powered section of the flight the XS-1 would glide back over Rogers Dry Lake at 400mph, having dumped any remaining fuel, drop the gear and flaps on a 250mph downwind leg and turn through 180° before making a final approach at 220mph – at its empty weight and at runway level,

the X-1 stalled at about 190mph. The aircraft, having no brakes, usually took about three miles to stop after landing.

In August 1947 Yeager made his first powered flight in the XS-1, and the AAF programme leaders realised that henceforth flight testing was going to progress more rapidly. Yeager dropped from the B-29 at 24,000ft, fired the four motor chambers in rapid sequence, pulled up the nose and rolled. As the aircraft went through zero-"g" the LOX tank cavitated and the XLR-11 cut. Yeager shut down the motor, dived to regain speed and pulled up before igniting all four chambers. The XS-1 bounded forwards, and to keep the aircraft within the speed envelope for the flight – Mach 0·82, below the onset of buffet – Yeager had to pull into a near-vertical climb. Observers watched in astonishment and alarm as the XS-1 blasted towards 40,000 feet in a vertical modified barrel roll and dived through Mach 0·84. The AAF "was a little upset", Yeager recalled.

Yeager pushed the XS-1 steadily harder through August and September. The thin-wing aircraft encountered buffeting at higher Mach than the No 2 machine (0·88, compared with 0·84) and beyond that point lateral and directional stability began to worsen as shock patterns affected the ailerons and rudder. Just before the fuel ran out on each flight Yeager would roll and pull 2-3 "g", to get some idea of the problems he would encounter at higher Mach on the next flight. When he did this at 0·94 he lost longitudinal control: it was found that the shock wave on the tailplane coincided with the elevator hinge line. Pitch control, as noted above, was probably the worst of the stability and control problems associated with high Mach numbers. The solution was to use the powered variable-incidence tailplane for control beyond the point where the elevator stopped working. Once Yeager had successfully performed a 3 "g" manoeuvre at Mach 0·94 indicated, he recalled, "as far as I was concerned we had got the thing licked". This was on October 10, 1947.

Muroc had not seen the first aviation activity in that particular section of the desert, this distinction being claimed by a fly-in restaurant, guest house and riding club owned by Florence "Pancho" Barnes, a former holder of the women's world air speed record. Riding at night was a favourite sport of the test pilots, and as the day of the planned supersonic run approached Yeager suffered a riding accident and broke two ribs. He could still fly the aircraft, but the problem was closing and latching the cockpit cover; a last-minute improvisation based on a broom handle overcame the difficulty. On October 14, 1947, Yeager was dropped from the B-29 and following a steep climb attained Mach 1·06 at 43,000ft. Beyond the mysterious barrier the buffeting stopped and Yeager recovered the use of the elevators. Eight miles below, observers heard the sound of the first sonic boom made by a manned aircraft.

In March 1948 Yeager took the XS-1 to 957mph, or Mach 1·4. (These figures, of course, did not qualify as official speed records, being strictly one-way runs, with an aircraft launched in mid-air, and taking place at high altitude.) In April George Welch of North American dived the XP-86 prototype – which also had a powered trimmable tailplane – through Mach 1, a flight which would probably not have been attempted without XS-1 experience. Later production Sabres were fitted with a "flying tail". This was a stabiliser in which the entire surface moved for control as well as for trim, the elevators being used to augment control at low speeds. Thanks to the XS-1 programme, the US industry was well ahead of its rivals in applying powered control to the pitch axis, and the difference was to prove important. While the Sabre's characteristics at high Mach were well known, and the aircraft could be dived past Mach 1 by ordinary service pilots, the same was not true of its adversary, the MiG-15. The Russian fighter's airbrakes were set to open automatically at Mach 0·92 to prevent it from reaching a speed at which it lost its elevator control.

Building an aircraft which could take off under its own power and go supersonic in level flight was now a matter of incorporating the XS-1 lessons into a production aircraft, and of providing sufficient power and fuel to do the job. The unaided turbojet engine was found to be inadequate to the task, at least at that time; a supersonic aircraft powered by a straight turbojet would have been all engine and little else. The answer was to employ afterburning, the system originated in Britain during the war years. The great advantage of afterburning was that it added a great deal of thrust for little added weight. The disadvantage was that it increased fuel consumption dramatically, because an engine in reheat produced much more thrust much less efficiently. This remains true to this day, and a military pilot who employs afterburner too extravagantly is apt to find himself gliding.

Afterburner design remained on the back burner (to coin a phrase) until the XS-1 showed that speeds could be attained where it would be useful. The only serious post-war work was done by Westinghouse, which had developed its own line of small axial jets for the US Navy and had fitted an afterburning J34 to the Chance Vought F6U-1 Pirate in April 1948. At that time the newly renamed US Air Force was undergoing a small revolution in its fighter philosophy. This had been brought about by a change in the service's attitude to the B-36 bomber. Generally considered to be a lumbering anachronism in the first few years of its development, the B-36 had demonstrated surprising performance at high altitudes, especially when fitted with auxiliary jet engines. The Air Force had concentrated on developing general-purpose fighter-bombers in the P-51 mould, but even the best of these – the F-86 Sabre – lacked the performance and stability at high altitude that the US Air Force would need to inter-

cept a Soviet counterpart of the B-36. The USAF began to concentrate more and more on high flight performance in its aircraft, at the same time authorising development of a new breed of specialised interceptors – the first USAF aircraft to be designed specifically for air defence since the XP-38.

A prototype level-supersonic interceptor had in fact been ordered as early as the end of 1945; after a very long development period it became the first combat-type aircraft to exceed the speed of sound in level flight. This was the extraordinary Republic XP-91. At the time it was designed, the most powerful engine available or in prospect was the GE TG-190 (later J47) and with the limited knowledge of reheat at the time this was considered inadequate to achieve supersonic performance or the high rate of climb that the USAAF was looking for. The XP-91 was therefore schemed as the first of a great many post-war "mixed-power" jet-plus-rocket interceptors. The idea had originated with the Me262C *Heimatschützer,* a rocket-boosted version of the German jet fighter that was intended to combine the fast climb of the Me163 with the jet's practicality, and underwent many variations ranging from mainly-rocket aircraft with a small get-you-home jet engine to a conventional jet fighter with rocket boost. The Republic XP-91 fell between the two extremes, with a reheated J47 and a Reaction Motors XLR-11 like that of the XS-1. Its aerodynamics were unconventional in the extreme. Like most designers in the mid-to-late-1940s, the XP-91 team was frightened to death of tip-stall. Not only was the wing slatted, but instead of simply eliminating or reducing taper the designers reversed it, so that the tips had greater chord and depth than the roots. The powered tailplane got the same treatment. As though this were not enough, the designers gave the wing variable incidence, so that the aircraft could fly slowly with the wing at a high angle of attack and the fuselage level, giving the pilot a reasonable view.

Redesignated XF-91, the prototype flew in May 1949, and more than three years later, with the rocket engine installed, it became the first combat-type aircraft to exceed Mach 1 in level flight. Republic's proposed production version, the F-91A with a completely redesigned and uprated GE engine, did not win an Air Force order: by that time there were easier ways of achieving similar results, and these bore fruit simultaneously in the United States and the Soviet Union.

It was the latter country, considered by the West to be limited to copies of Western and German designs, that succeeded in attaining level-supersonic speed with two turbojet-powered aircraft before this feat was accomplished by a Western aircraft. Joseph Stalin had decreed that the Soviet aircraft industry was to strive for world leadership. Stalin made few friends but knew how to influence people: disgrace, Siberia or worse awaited any

Ordered in 1945, the Republic XF-91 became the first aircraft to exceed Mach 1 after a conventional take-off. Powered by an afterburning J47, plus a rocket engine similar to that of the XS-1, the XF-91 was full of unusual features, including a variable-incidence wing that was deeper, and longer in chord, at the tip than at the root (Smithsonian Institution)

designer who failed in his set task. The key to the Soviet successes of the early 1950s was the rapid progress made by jet engines. Arkhip Lyulka had begun work in 1946 on the VRD-5, aimed at the then staggering thrust of 10,000lb. This engine ran as the AL-5 in 1950, and in February 1951 it flew aboard the Lavochkin La-190 fighter prototype. The La-190 featured a highly swept and tapered wing with large fences to control its worst characteristics, and a bicycle undercarriage like a smaller version of that fitted to the B-47. The La-190 is claimed to have exceeded Mach 1 by a small margin during its flight tests, but its handling was too poor for service use.

Meanwhile the Mikulin bureau was working at the other end of the size scale, on an engine nearly as powerful as a J47 but so small that a pair of them could be installed side by side in a light day fighter. The Mikoyan/Gurevich bureau was working officially on the Lyulka-powered I-350, but was quietly proceeding in parallel with the I-360 with two Mikulin AM-5s and a T-tail. This crashed due to tail flutter on an early flight, but by that time the I-350 had been modified to the twin-AM-5 I-350M. It was this low-tailed aircraft which made the first successful supersonic diving flights by the type; the

reheated AM-5 was not available, but this was expected to confer level-supersonic capability. The Soviet command was so impressed that a version of the I-350M with reheat was ordered into production off the drawing board, as the MiG-19F, and early examples of this aircraft were attaining Mach 1·1 in level flight by early 1953. But the production decision had been premature; the engines were unreliable mechanically and tricky to handle, and there was only one hydraulic system, driven by the starboard engine. The elevators became progressively less useful as speed increased – the Soviet designers had not realised that the days of simple elevator control were past. The design was revised as the SM-9 prototype, flown in September 1953; Sergei Tumansky redesigned the engines completely, as the RD-9, a one-piece slab tail was adopted and a dual hydraulic system was installed. This became the Mach 1·4 MiG-19S.

The development of the MiG-19 had been extraordinarily similar to that of its US counterpart; it was ordered into development within a few months of the Russian fighter, flew between the appearance of the MiG-19F and the first flight of the developed MiG-19S and had a similarly difficult development. Its design had started as a logical offshoot of the

97

F-86 Sabre. Ray Rice and Edgar Schmued of North American began work in early 1949 on a level-sonic fighter retaining most of the Sabre's features but with a thinner, more highly swept wing and a great deal more power: in fact the new aircraft was designed around an engine yielding twice as much power as the afterburning versions of the J47. This was the Pratt & Whitney JT3: the Connecticut company had been late entrants to the world of jet engines, having started out with licence-built versions of the Rolls-Royce Nene and Tay, but was to catch up in no uncertain fashion with the upstart GE.

P&W had been one of the companies which approached aircraft gas turbines as a straightforward replacement for piston engines, and had thus started with turboprop engines which would rival the fuel efficiency of piston engines. Its first gas turbine, in fact, was a compound of gas turbine and piston engine; the PT-1 was based on a 1940 design by Andrew Kalitinsky of the Massachusetts Institute of Technology for what could be considered a highly supercharged diesel engine, except that the power was extracted from the supercharger/turbine shaft. Component testing for this engine led to a Navy contract in June 1945 for a simpler pure gas turbine, although still of very high pressure ratio. This was the PT-2 turboprop, which was eventually adopted by the USAF as the T34.

High pressure ratio in a gas turbine offered the same advantages as high compression ratio in a piston engine: greater power for a given size and weight of engine, and more efficient extraction of energy from the fuel. While the limit to compression ratio in a reciprocating engine is set by the point at which the fuel detonates before ignition, the gas turbine runs into other problems: the compressor blades are more likely to stall, the engine proves more difficult to start, and the increasing temperatures which go with higher pressures tend to destroy the turbine. Some of these problems can be alleviated by designing an engine with lots of compressor stages – the PT-2 had 13 stages and Pavlecka's Northrop Turbodyne had 18 – but this added a great deal of weight and led to problems with torsional vibration in the shaft. Another approach, however, took advantage of the fact that as the air was compressed through the engine, the volume reduced and the stages got smaller. These smaller stages could be made to run faster than the front compressor stages and the rear turbine stages could manage, because the blades were smaller and the mechanical loads lighter; they could thus do more work without stalling. By 1943 the RAE had worked out a practical method for doing this, designing a jet engine with two concentric shafts. The inner shaft carried the front (low-pressure) compressor stages and the rear (low-pressure) turbine, while the faster outer shaft carried the rear compressor stages, driven by a high-pressure turbine immediately downstream from the combustors.

Soviet designers espoused the conventional highly swept wing despite the potential low-speed problems. After a tricky development the MiG-19 emerged as the world's first level-supersonic fighter. Other advantages included a high rate of climb and a heavy armament

99

North American's F-100 was similar in many ways to the MiG-19, with reheated turbojet power and a slab tailplane. Both the wing and the tail were set low, and were thinner and less sharply swept than those of the MiG-19

Pratt & Whitney adopted this approach for the new turbojet engine which, after the PT-1 and PT-2 propeller-turbine engines, became the Jet Turbine 3 or JT-3. (The company later dropped the hyphen from these designations, which continued in an unbroken series until 1981.) Not only was the JT3 selected for the North American level-sonic fighter, but it found its way into Navy fighters, USAF strategic bombers and tankers and the first generation of US jet airliners. It was scaled up into the JT4 and scaled down into the JT8; both the JT3 and the JT8 led to turbofan derivatives that virtually monopolised the entire civil jet market until 1970, and retain a stranglehold on the market for jets under 160 seats to this day. The conclusion that P&W must have got something right is inescapable.

The aircraft that was to use the JT3 in its after-burning form – P&W had already designed a successful afterburning system for the Tay, so reheat held few terrors – gradually evolved from a hotted-up Sabre into a completely new machine. To avoid possible aileron reversal, the roll-control surfaces were moved inboard to replace the flaps. The nose was extended and at the same time made flatter to preserve the cockpit view, giving the nose inlet a distinctive shape which was to earn the aircraft the name "Satchmo" among some of its crews. Designated YF-100A, the prototype flew in May 1953, and immediately did something quite unprecedented: with reheat lit on its XJ57 engine, it exceeded Mach 1, on the level, on its first flight.

With Sabre experience behind it the F-100 Super Sabre had no problems with its pitch-axis control; the aircraft was designed with a one-piece slab tail. Roll and yaw control were problem areas. During development the F-100 fin was shortened to reduce drag, and this was followed by a series of unexplained high-speed accidents. It was found that the F-100 had introduced a completely new phenomenon to service aircraft: "inertia coupling". The fuselage of the F-100 was long and heavy in relation to the wings. If the aircraft encountered a critical combination of pitch and roll, gyroscopic forces might try to pull the fuselage into the plane of the roll. The aircraft would tumble violently and disintegrate. Finally, the problem was cured on the F-100 by adding wing and tail area, damping down the deadly forces enough for service use.

Eventually the modified Super Sabre and the modified MiG-19S reached the squadrons in the same year: 1955. They had had yet a third close contemporary in the Hawker P.1083, a development of the Hunter with the same key features as the American and Soviet fighters – reheat, a slab tail and a highly swept thin wing – but it was scrapped while nearing completion. It was the French industry that flew the first production supersonic fighter in Western Europe: the very F-100-like Dassault Mystère IV. At that time, there might even have been some optimists around who thought that the problems of supersonic flight had been solved. This was not so: the fun had hardly begun.

100

800–1400mph

The fertile Fifties

800–1,400mph

The fertile Fifties

ALL sorts of weird and unearthly shapes were being dreamed up in the early 1950s in pursuit of high supersonic speeds: shapes that promised altogether new and wonderful difficulties in control and stability, that presented challenges in construction or promised even higher unstick and landing speeds. Some of these odd-looking aircraft survived the prototype and production stages, and some are in service today: we have simply grown used to their eccentric appearance, but in the middle years of the century the imagination of the aircraft designers seemed to have outstripped that of the comic illustrators.

Alongside these aircraft, however, there proceeded an almost ignored stream of aircraft development based on the classic swept wing: less exciting, mind you, and possessing less visual impact than the types with which the decade is usually identified, but capable of remarkably similar performance.

In fact the first aircraft to attain Mach 2 was just such a design, with a simple, untapered, moderately swept wing based on German design data. This was the Douglas Skyrocket, developed under the auspices of the US Navy and NACA. As noted in the previous chapter, these two organisations had taken a different road to high-speed research from the path blazed by the US Army Air Force, preferring to develop research aircraft within the constraints set by unassisted take-off. The first of these research aircraft was the Skystreak, with a thin-section straight wing and a turbojet engine, but before the Skystreak set its 1947 speed records the development of a much faster aircraft was well under way. This combined a swept wing with a classically streamlined fuselage, smoothly curved from nose to tail and of quite considerable girth; the fuselage had to be big, because it was intended to accommodate not only a Reaction Motors LOX/alcohol rocket but a turbojet as well.

The aim was to build a highly supersonic aircraft that could be operated from a normal runway, but it soon became apparent that the old problems of endurance and weight still remained insoluble: the aircraft could not carry enough rocket fuel to attain its maximum potential altitude and speed, even with additional solid rocket boost for take-off and a three-mile run. Shortly after the first mixed-power flights, it was decided to convert the second Skyrocket for air-launched operations with added fuel capacity, and it was in this aircraft that Bill Bridgeman of Douglas attained Mach 1·88, or 1,260mph, in August 1951. (The aircraft had a major advantage over the early X-1, in that Reaction Motors

engineers had finally made a turbopump work and the hefty nitrogen system could be dispensed with.) After that the aircraft was handed over to NACA, and it was two years before it attained a similar speed again. Eventually, in November 1953, Scott Crossfield took the second Skyrocket to 1,291mph at 62,000ft, which at that altitude was equivalent to 2·005 times the speed of sound. But by that time a great many designers had concluded that only radically unusual shapes could reach such high speeds.

Like so many ideas in high-speed aerodynamics, the most popular of these new solutions had emerged from Germany. Investigating one of the Reich's many research establishments, a USAAF technical team found an aircraft that made the Ju287 look quite normal: it consisted of a triangular wing, with its leading edges swept back 60° to meet a straight trailing edge. A huge vertical fin of similar shape and proportions reached to the nose and housed the cockpit. This was the DM-1 glider, built to the orders of Alexander Lippisch (the designer of the Me163) to test what he called the "Delta" wing. Lippisch had planned to use the configuration for a supersonic interceptor, the P-13, which in view of Germany's shortage of fuel was designed to run on coal.

Lippisch believed that the delta, unlike an ordinary swept and tapered wing, would not be afflicted by pitch-up due to tip-stall. Also, he said, the wing would not stall in a conventional manner at all. As the nose went up the wing would keep producing lift, but at steadily decreasing efficiency. Instead of stalling the delta would just run out of power to overcome increasing drag and would sink vertically. The long trailing edge provided room for powerful and effective control surfaces, and was so far aft of the centre of gravity that the same surfaces could be used for both pitch and roll control. The wing would have such long chord that it could be made aerodynamically thin – expressed as a percentage of the chord – while being structurally thick enough to house fuel and landing gear. The detractors of the delta argued that it would have similar tip-stall characteristics to a swept wing, only worse, and that without a tailplane to provide damping the aircraft would be uncontrollable. It was impossible for both sides to be right, and the complex aerodynamic factors involved, combined with the newness of the configuration, meant that full-scale flight test was the only way to find out.

By 1946 two efforts in this direction were under way. The British were working on slightly less swept variants of the delta configuration. Roy Chadwick at Avro was starting design of a high-altitude strategic bomber which blended Northrop and Lippisch ideas, being a thick-section flying wing of 45° delta planform; this became the Vulcan after gaining a few vestiges of a fuselage in the design process. This line of development was firmly aimed at high subsonic speeds. The USA was aiming at higher performance targets. As noted in the last chapter,

First aircraft to Mach 2 was the fairly conventional Douglas Skyrocket. Sponsored by NACA and the US Navy, like the Skystreak, it proved that non-exotic configurations could attain high speeds. The aircraft is shown here on an early test flight, powered by a jet engine. This was later removed to make room for more rocket fuel, and the aircraft was air-launched for its Mach 2 flights *(Smithsonian Institution)*

the USAF had issued a requirement for a supersonic short-range interceptor towards the end of 1945, one of these aircraft eventually appearing as the Republic XF-91. A parallel contract went to Convair, despite the company's relative lack of fighter experience, and covered a swept-wing aircraft designed by a specially formed fighter team. Tunnel tests in 1946 proved that the original XP-92 design would have unacceptable pitch-up and lateral control problems; in the same month that the Convair team began tests of a 60° triangular wing, they encountered Lippisch's work. Convair's main innovation was to use the delta as a means to attain very thin wing sections without structural problems. The DM-1 glider had a 15 per cent thick wing, but the redesigned Convair XP-92 had a wing of 6·5 per cent thickness. It also possessed a unique compound ramjet/rocket engine and was designed to take off from a trolley. Roll and pitch control were to be combined in one large surface on each wing for which the portmanteau word "elevon" was coined. Too large to be operated manually, these surfaces were driven by irreversible hydraulic jacks.

Late in 1946 the USAF and Convair decided that the XP-92 design possessed so many new features that it would be wise to develop a cheap low-speed test aircraft to check out the aerodynamics and controls: this would be to the XP-92 what the Ju287 V1 was to the production aircraft. The wings and control system of the test aircraft, the XP-92A, were to be similar to those of the fighter, but it was to have a normal landing gear, a turbojet engine and a conventional canopy and ejection seat for the pilot.

The XP-92A was completed in late 1947, and was one of the first important aircraft to be tested in NACA's most expensive new facility: a colossal new low-speed wind-tunnel, with a test section measuring 40ft by 80ft, built at the former Navy airship base at Moffett Field, California, and specifically intended to test full-scale models and actual aircraft. The tests failed to vindicate the detractors of the delta, and the XP-92A was transferred to Muroc in April 1948. Taxi trials and cautious hops off the ground showed that the aircraft was reasonably stable as long as the pilot did not try to do anything, but that the big powered elevons had dramatic and unpredictable effects on the flightpath. Before a full flight could be attempted, the US Air Force scrapped the XF-92 fighter – the equations of thrust, drag, weight and endurance were once again producing the wrong answers – but reprieved the XF-92A as a pure research programme. The control bugs were slowly ironed out, as pilots gained a better idea of coping with the elevon forces, and the powered control system was refined and made less hypersensitive. In September 1948 the XF-92A became the first delta to fly, and in the following year Chuck Yeager and Pete Everest not only dived the aircraft through the sound barrier but flew it as slowly as 67mph with the nose at a crazy 44° skyward angle.

These slow flights, however, did not mean that the delta was a road to short take-off and landing: far from it. At practical angles of attack the delta behaved like any other short-span wing, with very high induced drag and poor lift. Worse, it was impossible to put flaps on a tailless delta, because there was no tail to trim out the nose-down pitch caused by the flaps, while the elevons were so close to the centre of gravity that they needed to generate a very large download to lift the nose; they did not have the leverage of the conventional tail. All this meant that the delta had to land at high speed, and two-mile runways began to be standard equipment for military airfields. (Less than ten years before the XP-92A flew, the majority of the world's military aircraft could operate quite easily from a grass-covered arena. The minority which could not do so were flying boats.)

By mid-1951, the XP-92A had proved successful enough to form the basis for a new and more practical fighter, made possible by the development of more advanced jet engines. The new aircraft was ordered into production while still in mock-up form, to save money and time; it was designed to defend America against Russian nuclear bombers, and was to be missile-armed and fully automated as well as supersonic. The decision to carry missiles internally to reduce drag, plus the need for considerable internal fuel capacity, endowed the Convair Model 8, or YF-102, with a somewhat stout fuselage, which was very nearly its undoing. When the prototype was flown in late 1953 – with production aircraft rolling down the line behind it – it refused to attain Mach 1, a doubly humiliating failure in view of the ease with which the more conventional and similarly powered Super Sabre had broken the barrier on its maiden flight. The problem was the combination of thick delta wing and short, tubby fuselage, which caused sharp changes in the cross-sectional area of the aircraft from nose to tail. These in turn generated rapid increases in local air velocity and, at critical Mach numbers, strong shock waves. It was not entirely unexpected: before the F-102 made its first flight NACA aerodynamicist Richard Whitcomb completed an investigation of the theoretical basis of the problem, and formulated his Area Rule; this, basically, warned against drastic changes in cross-sectional area, relating wave drag to such changes.

A 16ft fuselage stretch and silicone implants saved the F-102A programme, while Convair incorporated area-ruling more tidily in the new and faster F-102B – to emerge as the F-106A – and in its very ambitious XB-58 supersonic medium bomber. Whitcomb was virtually canonised, and for a time some degree of "wasp-waisting" seemed mandatory on any US design. But quite a number of aircraft had meanwhile been designed in total ignorance of Whitcomb's research, and they included aircraft substantially more efficient than the Convair deltas.

Two of these were designed in Britain, whose industry was making a heroic catch-up effort after

being left behind in axial engines and swept wings. Just two years after the final cancellation of the Miles M.52 in 1946, the RAE established a project group aimed at developing a controllable, efficient, high-supersonic fighter configuration. The delta seemed attractive – the chairman of the group, Morien Morgan, went on to design the modified-delta wing of Concorde – but there was also a strong argument for retaining a tail for control, damping and stability. The solution to the argument adopted in 1949 was to order two high-speed research aircraft to specification ER.103, one with a pure 60° delta wing and one with a tailed configuration. Both were turbojet-powered aircraft and were designed to form the basis of fighter designs, and in a sense both achieved that goal. Neither, interestingly enough, stemmed from a design team with any strong fighter background.

Fairey was to build the delta design, having been working on subsonic deltas since 1947 under a programme aimed at producing a ramp-launched target-defence interceptor. The prototype for this aircraft was the tiny Fairey Delta 1, flown in 1951, so

the supersonic aircraft became the Fairey Delta 2 or FD.2. A contemporary of the YF-102, it did not have to fit so much inside its fuselage and, although the Fairey designers were not aware of the Area Rule, they knew enough to keep the cross-section constant and small. Like the Convair deltas, the FD.2 had powered controls with no provision for the pilot to take over manual control, but it differed in having separate ailerons and elevators. Its wing was thinner and its fuselage more slender than those of the American aircraft, and in addition it had one remarkable gadget: a hinged nose, dubbed a "droop-snoot", which could be raised for optimum streamlining, or lowered for take-off and landing to give the pilot a reasonable view.

While the FD.2 bore a close resemblance to the Convair family, the other aircraft designed to ER.103 was unique. The English Electric P.1 – designed by a newly formed team whose only previous aircraft had been the Canberra bomber – had what amounted to a 60° delta wing with a large triangular bite removed from the trailing edge. Although this threw away the structural advantages

To this day the wing shape which W. E. W. "Teddy" Petter selected for the P.1 remains unique, combining many of the separate advantages of highly swept and delta wings. Also of interest is the fuselage design: while other designers were drawing bullets, Petter hit on the advantages of a tube (BAe/English Electric)

of the delta, it retained its tip shape and made room for unswept ailerons, avoiding many of the problems of tip-stall and lost aileron effectiveness that plagued the design of conventional highly swept wings. The big advantage of the P.1 wing, however, was that there was room for a conventional tail on the rear fuselage, and for flaps on the trailing edge of the wing, so that the wing could be smaller and more efficient than a straight delta.

In the absence of the area rule, Teddy Petter (designer of the advanced pre-war Westland Whirlwind) played safe and designed the P.1 with a fuselage that had as much cross-sectional variety as a length of drainpipe. Recognising that inlet and exhaust flows took up less cross-sectional area than the complete engine, he installed two engines, one ahead of and below the other, so that the cross-section of the twin-engined P.1's fuselage could be less than two engines side by side would dictate. Petter left English Electric in 1950 to pursue his own lightweight fighter ideas, leaving the P.1 in the hands of Frederick W. Page (later Sir Frederick Page).

Following the American example (and, indirectly, that of the Ju287), the British tested the P.1 configuration on a low-speed prototype, the Shorts SB.5. Not only could the wing sweep of the SB.5 be changed over a small range on the ground, but the tailplane could be installed either on the tip of the fin – favoured by the RAE – or on the base of the fuselage as the English Electric team preferred. (The F-100 was under construction with a low tail at that time, reversing years of fighter practice.) The low tail, built as a simple mechanically continuous slab, was selected for the P.1.

Both British aircraft flew in 1954, and in August the P.1 became the nation's first aircraft to go supersonic in level flight. The FD.2 programme was delayed by an accident at an early stage, and it was October 1955 before it equalled the P.1's feat. It was quickly clear that the FD.2 was an extremely fast aircraft. The Fédération Aéronautique Internationale had just introduced new rules for speed record attempts at high altitudes, and the USA had promptly set the first speed mark in the new category with an F-100A, at 822mph. The FD.2 could

106

take the record with no trouble, and preparations were secretly made for a record attempt. The secrecy, plus the fact that the development of the FD.2 had aroused little interest outside the UK, promised a royal surprise for the Americans and a morale-booster for the British industry. In March 1956 Peter Twiss bettered most expectations with a record-shattering 1,132mph run over the South Coast. This was 37 per cent faster than the F-100, and the biggest percentage increase in history.

It was the P.1 which was chosen for further development as Britain's first supersonic fighter, and it became the first British aircraft to attain Mach 2. In its service form it was probably the first twin-engined aircraft to go supersonic on one engine, and the first to go supersonic without reheat. Meanwhile the FD.2 provided inspiration and some information for the very successful Dassault Mirage III family. The developed service version of the P.1, the Lightning, proved to be an interesting application of the Area Rule: according to the rule, the designers discovered, it should be possible to increase the cross-section of the Lightning's rear fuselage without increasing transonic drag. This proved to be the case, and the fighter acquired a sort of kangaroo pouch under the fuselage, which steadily grew in size as the aircraft developed through successive versions and did much to rectify the Lightning's basic problem: its lack of internal fuel capacity. The engines also presented a point of significant interest: while the original P.1 had been fitted with Sapphires – which engine, originally run as the Metrovick F.9, was the first of the RAE/Metrovick series of engines to enter production – the production Lightning had Rolls-Royce Avons of the new RA.24 type, in which permissible operating temperatures were increased through the re-introduction of air-cooled turbine blades. Rolls-Royce was a pioneer in applying this old German technique to modern cast blades.

Another design that antedated the Area Rule, and emerged none the worse for it, resulted from a requirement issued in 1949 by the Royal Swedish Air Force, calling for a supersonic interceptor. For a country which had barely possessed an aircraft industry before the wartime interruption of imports had forced it to do so, this was audacious in the extreme. The design which the Svenska Aeroplan AB – usually abbreviated to Saab – submitted to meet the requirement was even more remarkable, being a flying-wing delta with 70° of sweepback. Tunnel tests showed the aircraft to have horrible low-speed characteristics, but instead of scrapping the concept completely the Swedish designers extended the span by adding thinner-section, less sharply swept outer panels: the so-called "double-delta" configuration. This was extensively tested on the little Saab 210 research aircraft (it spanned only 16ft) three years before the prototype Saab J35 Draken fighter flew. It proved an extraordinarily successful compromise, offering huge internal volume for low drag and hence permitting a very compact and light structural design.

Two more designs fall within a review of early supersonic configurations, both stemming from the United States, and both ordered into development in the first half of 1953. The first to see the light of day was a Navy fighter, with the same engine as the F-100 and a very similar wing – of thin section and moderate sweepback, and with the ailerons located well inboard, precluding the use of flaps but making it easier to avoid twisting and control reversal. Where the Chance Vought F8U Crusader differed from the F-100 was in its fuselage, which was long and untapered like that of the P.1, and where it differed from nearly all of its contemporaries was in the junction of its wing and fuselage. The wing could be mechanically tilted upwards by 7° – or it might be more correct to say that the fuselage could be tilted downwards, since the object of the exercise was to give the Crusader a short and simple landing gear, together with a good view for the pilot.

The Crusader had a mixed reputation. On the one hand it was probably the fastest fighter of its day, the early versions exceeding 1,000mph for the first time in a service aircraft, and in later life it proved to be one of the few early US supersonic fighters that had retained the ability to manoeuvre in combat. On the other hand it had one of the highest sustained accident rates of any aircraft built; by the late 1950s, it was reckoned, a US Navy pilot had a 23 per cent chance of being killed during his 20-year flying career if the accident rates then current were maintained, and the same pilot had a better-than-even chance of being forced to eject from his aircraft.

The Crusader, however, was always overshadowed in controversy by its closest contemporary. While regular reports were received on progress with such advanced aircraft as the Crusader, F-102 or P.1 in 1954–56, there were constant rumours of wholly bizarre shapes appearing in the sky above California, and of strange things taking shape under a near-impenetrable cloud of secrecy in Lockheed's Burbank plant. Then the FD.2 resoundingly beat the US-held speed record, and the USAF ripped the veil aside and showed to a bemused world the Lockheed Model 83, or F-104 Starfighter.

The F-104 was a product of events: the shift in US fighter priorities following the 1948 re-evaluation of the bomber threat, the explosion of the Russian atomic bomb in 1949, and air combats over Korea. The former events caused the cancellation of Lockheed's XF-90, which carried with it a great many of the company's hopes for a successor to the F-80 Shooting Star. When Lockheed went out to discover what the fighter pilots of the Korean War wanted, they heard complaints about Soviet fighters that could outclimb the best the USAF had and match the Sabre in level speed. Kelly Johnson decided to build what the pilots wanted – an uncompromised performance fighter, carrying one powerful gun and

a pair of the new infra-red homing missiles then under development for the US Navy.

Johnson hand-picked a small team to design and build the new aircraft, just as he had done with the XP-80 in 1943. Within the design team there were no organisational charts; flexibility was kept to a maximum and the specialists in each area of design aired their problems in open session. Team members understood the need for secrecy, and shared this understanding even with their former colleagues in other Lockheed divisions. Rank alone bought no admission to the stark beige hangars or the offices where Johnson's team was at work. Somebody likened it to the noxious-smelling factory in the comic strip *Li'l Abner*, where the mountain town's moonshiners brewed Essence of Skunk, the final touch to their secret recipe for "white lightnin'". And so the nondescript clutch of buildings housing the advanced project team became the Skunk Works, and the little black-and-white animal with a strong sense of privacy and a short and horrible way with those who violate its space became the symbol of the world's most forward-thinking and effective aerospace design team.

Ultimate performance for the new fighter, which the team began to study in 1951, meant low drag and high power. Kelly Johnson recognised that the best solution to the problem of supersonic lift/drag ratio was a straight wing of very thin section. Having only a tiny leading-edge radius, it would generate only a weak shock wave, while its camber would be too small to create other shocks. If the section could be kept thin enough, there would be no need for the weight and complexity of sweep-back. The snag was that the wing had to be considerably less than five per cent thick, and the only way to produce a straight wing of such thin section that would still be strong and rigid was to make it extremely short.

Very thin, straight wings are in many ways an aerodynamic ideal for supersonic flight. The problem is that they have to be very short to be structurally practicable. They first appeared on the highly ambitious Douglas X-3, ordered in 1945 for sustained Mach 2 research *(Smithsonian Institution)*

Kelly Johnson's original XF-104 Starfighter was a lightweight machine with advanced but simple systems. Like the XP-80, it was redesigned around a bigger General Electric engine after flying with a British powerplant. Note the sharp-lipped semicircular inlets, replaced on the YF-104 by half-cone types *(Smithsonian Institution)*

An aircraft based on this principle was in an advanced stage of construction a few miles from Johnson's design offices as work on the Lockheed fighter started. This was the Douglas Model 499D, or X-3 Stiletto; designed to an ambitious mid-1945 requirement for an aircraft capable of attaining Mach 2 for up to 15min after a conventional take-off, the Stiletto featured a 66ft-long, dagger-shaped fuselage supported by tiny 22ft-span unswept wings of 4·5 per cent thickness. Even with flaps on the leading and trailing edges of the wing, the Stiletto was not expected to unstick at less than 244mph. Engine trouble – specifically, the failure of the Westinghouse J46 to appear on schedule – was well on its way to wrecking the entire programme by 1951. However, all NACA's wind-tunnel data on the X-3

– in particular, on its dangerous inertia-coupling characteristics – were made available to Lockheed.

Engineering design of the Model 83 started in late 1952, and early in the following year the Air Force ordered two XF-104 prototypes. The first flew just one year later. The wings of the new fighter spanned about the same as those of the X-3, but were sharply drooped to improve stability and even thinner than those of the research aircraft. The actuators for the trailing-edge and leading-edge flaps were inside the fuselage; the ailerons were powered by tiny multiple actuators buried in the wing. In an effort to make the XF-104 land at something like a normal speed, it was fitted with so-called Attinello flaps in which a high-pressure jet of air from the engines was blown through a slot ahead of the flap, allowing it to be

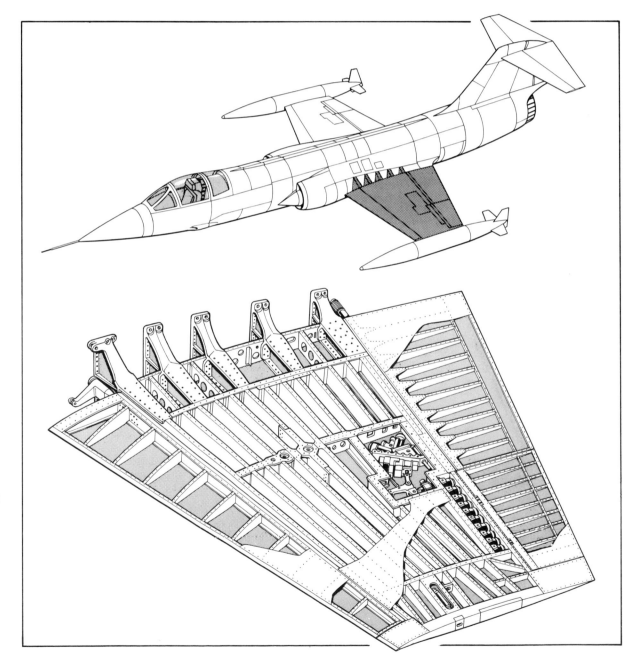

The thin, straight wing of the Lockheed F-104 was designed without ribs. Instead, the skin was built up on a series of spanwise forged spars. The movable leading-edge and trailing-edge flaps could be actuated from inside the fuselage, but the control mechanism for the ailerons had to be buried in a cavity within the almost solid structure. There was no room for a normal actuator, so each aileron was moved by a "piccolo" arrangement of ten tiny jacks.

sharply drooped without flow separation. These blown flaps were to distinguish a number of aircraft – particularly carrier-based types – during the late 1950s and 1960s. Most of the XF-104, however, was housed in its fuselage: engine, landing gear, the big multi-barrel T-171 gun and all subsystems. The powerplant of the XF-104 was the Wright J65 – none other than the British Sapphire of RAE/Metrovick ancestry – with which the Lockheed fighter was closely comparable in speed to the FD.2.

Already, however, the lightweight F-104 was being considerably redesigned for production. The key to the larger and heavier YF-104 was a new engine, which was to push the Lockheed aircraft past Mach 2 and would eventually power more Mach 2 aircraft than all other Western engines put together. Like the Pratt & Whitney J57 it was produced by a company which had seen its leading position in the market rapidly eroded: this time it was General Electric which had suffered. In 1947 it seemed that nearly every new or projected US aircraft had the J47: five years later GE had seen its rival's J57 selected for a huge range of aircraft, while the other old-established engine builders, Allison (which had got its original jets from GE) and Wright (with the licence-built Sapphire), were also cutting slices out of the market. GE's planning had gone awry. Its first follow-on to the J47, the XJ53, was much too big, eventually delivering nearly 24,000lb thrust with reheat, which was far more than anybody needed. Then GE seemed to over-react, designing a high-pressure-ratio, high-efficiency engine planned to fit in the same airframes as the J47. Originally known as the J47-GE-21 and later redesignated J73, it was too small for most fighter projects and was used only in a few F-86H Sabres.

The J73 had embodied a number of new features, including movable vanes to guide and regulate the flow into the engine. In particular, these could be closed to reduce airflow and avoid starting problems with the high pressure ratio of the J73. In 1951 GE began the design of an engine in which this principle was extended to the stator blades between the rotating compressor stages. Using variable stators, a single-spool engine could be made to behave properly at a high pressure ratio, and to perform efficiently over a very wide speed range. At high speeds, for example, the stators could be adjusted to admit only as much air as the higher-pressure rear stages could handle, preventing the development of stalls at the back of the compressor. In a sense, the stators acted on the rotor blades like flaps or slats on a wing.

Tests of a variable-stator compressor had achieved such promising results by late 1952 that GE dropped its parallel work on two-spool engines. Work started on the GOL-1590, a variable-stator experimental engine for ground tests, and the similarly sized MX-2118, aimed at the supersonic bomber which Convair was developing for the USAF. The latter engine was ordered into full-scale development in early 1953, and was designated J79.

The J79 gave about as much static thrust as the J57, but was in other respects very different. It had a 17-stage compressor, which made it a long and skinny engine, suited to fast and slender aircraft. (It had to be long and thin, because the supersonic Convair bomber was to have podded engines, and fat engines would have made compliance with the Area Rule impossible.) The first six stages of the compressor featured variable stators. The turbine blades used the first of GE's newly developed Rene series of high-temperature nickel-based alloys.

Also, GE worked on the design of the new engine in very close contact with the airframe designers. By this time it was becoming clear that the engine installation, and in particular the shape of the intake, had a strong influence on high-speed performance. The first jet engines had been fed with air through simple forward-facing holes, with scoops if they were needed to scrape off the slow and turbulent boundary layer around the fuselage. Beyond Mach 1, however, the shock waves thrown off by the intake lips generated increasing drag and inefficiency. In theory it was possible to design an intake with a converging duct so that the shock wave would fall completely within the inlet, and the energy expended in compressing the air would be recovered as thrust. The problem was that such intakes were hard to design in the first place, and would then only work at one given Mach number. A simpler palliative was to design the inlet so that just upstream of it there was a "compression surface" – an object that would trigger a series of weak shock waves rather than a single strong one, allowing some pressure recovery under controllable conditions. This could be designed to generate either a two-dimensional flow pattern, as in the case of a ramp or a wedge with a rectangular inlet, or a three-dimensional pattern, as in the case of a circular or semi-circular inlet with a cone or half-cone in the centre.

Half-cone inlets were fitted to the J79-powered development of the Starfighter, the YF-104A. They were considered so secret that they were actually covered with aluminium panels when the USAF whisked the new fighter out of its hat in early 1956.

The Starfighters proved extremely fast and more than a little tricky to handle. Part of the trouble was that the radical J79 had little flying time when the F-104 started flight tests; in a deadstick landing, the tiny wing with its blown flaps inoperative was not a great deal of use. But it was probably one of the first supersonic aircraft in which the maximum speed was limited by factors other than power: pressures and temperatures within the intake and on the airframe kept its service speed to a nominal Mach 2·2. It was also extremely fast at low level; a modified Starfighter currently holds the world's air speed record under the old low-level regulations, at over 1,000mph.

Even the Starfighter's technology began to look a

little dated by the late 1950s, particularly in the area of the engine installation. Although the J79 reigned supreme as a high-speed powerplant, it was, in the case of the Lockheed fighter, installed behind an inlet which was to some extent a compromise design, intended to give acceptable performance at all speeds. In the later 1950s, however, a number of aircraft began to appear with movable inlets and exhaust nozzles, which could adjust to suit different conditions. One of the simplest systems was that of the Convair B-58; the cones in the circular inlets moved forward at high Mach numbers, spilling the primary shock wave past the inlet and keeping flow roughly constant. Antonio Ferri of NACA created one of the first inlets designed to hold a stable internal shock wave. The outer lip of the Ferri inlet was extended well forward, giving the appearance of a gaping mouth. The forward lip generated a primary shock wave that was free to spill overboard in the V-shaped gap between the lip and the fuselage. Just inside the inlet, on the opposite side from the extended lip, a pair of hydraulically powered ramps moved inwards to narrow the inlet throat and at the same time generate a second set of shock waves, compressing the incoming air and effectively "supercharging" the engine.

The raked-forward Ferri inlet was used on only two or three aircraft, and only the Republic F-105 went into production with it. A closely related design, however, was adopted by North American for the experimental F-107, which had started development as an advanced version of the Super Sabre and emerged in 1956 as a totally different aircraft. The F-107's single engine – a P&W J75, the uprated and modernised descendant of the J57 – was fed by a double inlet on top of the fuselage, directly behind the canopy. Each side of the inlet resembled a Ferri type in which the forward lip had been moved to the aircraft centre-line, on the same side as the internal ramps, creating a wedge shape. On the F-107, two of these inlets were installed side by side; but these modified Ferri intakes could be used singly and turned through 90° so that the ramps moved in a vertical plane.

It was these inlets that were fitted to the most technically ambitious and advanced supersonic aircraft of the 1950s. Like so many advanced aircraft, it stemmed from a newly formed design team, headed by Frank Compton at North American's Columbus division. Columbus had been acquired from Curtiss-Wright as a production facility for Navy aircraft, and Compton's team had a natural interest in what was seen as the Navy's most important combat aircraft: the nuclear bomber that made the USN's CVA-class carriers into strategic weapons. North American's AJ Savage had been the first aircraft in the class, but was to be replaced by Ed Heinemann's Douglas Skywarrior. NAA's studies of a replacement for the Skywarrior in the nuclear attack role started as a low-level, subsonic aircraft designed to fly "under the radar". In early 1955,

Chance Vought's Crusader showed what could be done with a wing of moderate sweep and thin section. That this would work even at very high Mach was demonstrated by the 1,550mph F8U-3 Crusader III. The Jaws-style inlet was built to one of the earliest variable-ramp designs, by NACA's Antonio Ferri (Smithsonian Institution)

112

Aircraft such as the Rockwell Vigilante ushered in the era in which the engine maker and aircraft designer have to work together at the preliminary design stage, so that the engine can function as efficiently as possible behind the variable inlet. Detail (a) shows the inlet at subsonic speed, with a smoothly curved throat. Detail (b) shows the ramps of the modified Ferri inlet extended to narrow the throat and create a secondary shock wave — the primary shock has been spilled past the inlet by the extended top lip — which compresses the air before it is fed to the engine. Bleed air is allowed to escape between the ramps to stabilise the flow, though at some cost in efficiency. Note also the exhaust iris closed at low speed (a) for low drag, and open (b) for maximum thrust

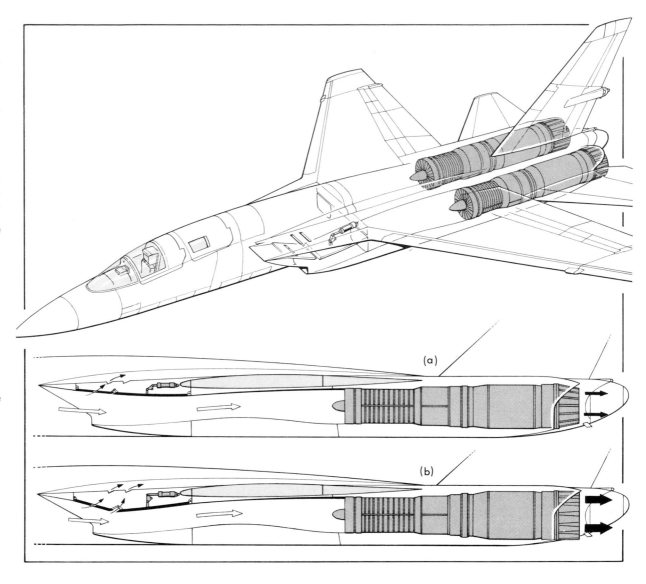

(a)

(b)

The elegant Vigilante was the most advanced aircraft of 1958. Its inlets were to become standard on later supersonic aircraft, from the Concorde to the F-15 Eagle, while its thin-section shoulder wing was to be echoed on many other aircraft. It also provided the inspiration for the MiG-25 Foxbat

however, the Navy and the Department of Defense insisted that the aircraft should be capable of Mach 2 at high altitude and should be able to take off at maximum gross weight from a moored carrier in a flat calm. (The North American bomber was the only aircraft ever designed to meet the latter requirement, which was rescinded before it entered service.)

Challenged to undertake this virtually impossible task, Compton and his team nevertheless proceeded rapidly enough to be given a go-ahead to build two YA3J-1 prototypes in mid-1956. Experience from all over the North American organisation went into the design. The wing had to be of considerable area and span thanks to the stiff take-off requirement, but had to be thin for minimum supersonic drag. Eventually, a 35° swept wing, 3·5 per cent thick, was chosen. To achieve adequate strength, the skin and the spanwise stiffening members were to be carved from solid plates of new aluminium-lithium alloys by automated machine tools. The engine bays used

a great deal of titanium, probably the first time so much of the new material had been designed into one airframe. Parts of the aircraft were almost literally gold-plated: heat-reflecting gold film was used in the engine bays.

The whole span of the wing was needed for blown flaps, so the ailerons were dispensed with and replaced by spoilers – a technique pioneered by Northrop in the wartime Black Widow. For the first time, the slab tail was made in two independent sections, which could be moved differentially to trim the aircraft in roll.

The inlets for the J79 engines were vertical wedges, and the exhaust nozzles were made variable in shape and area for efficiency at low and high speeds, and at different reheat settings. Directional control promised to present problems, because a normal rudder could be blanketed by airflow from the wide fuselage at high angles of attack, so the original A3J design was given two outward-canted tailfins. Before the prototype was completed, how-

ever, these were replaced by a single all-moving slab fin.

The size and power of the slab tail surfaces were unprecedented, and the control system that drove them drew heavily on NAA Autonetics' experience with the Mach 3 Navaho cruise missile. Control commands were translated into electronic signals which activated the hydraulic jacks powering the fin and "tailerons", although conventional mechanical signalling to the jacks could be used if that failed; this was the first application of "fly-by-wire" technology, although the term itself would not come into general use for more than a decade. With the power and sensitivity of the controls, the automatic flight control system – as the autopilot had been renamed – assumed greater importance. That of the A3J was designed to soften pitch characteristics, which were expected to be tricky at high speeds. At low speeds the tailerons moved in proportion to the movement of the stick (as they would if connected by cables); as speed increased the AFCS switched into a "'g'-command" operating mode, translating force on the stick into control deflections to give a proportional pitching moment. The changeover was programmed to take place gradually between Mach 0·25 and Mach 0·55. This was one of the first stability augmentation systems (SAS) in the highly sensitive and crucial pitch axis to be designed into any aircraft.

The first YA3J-1, named Vigilante, flew in August 1958 and embarked on an arduous flight-test programme. There were a lot of sophisticated systems to be developed, and many of them carried out the most crucial functions of flight control. In service, one pilot recalled, "it was not an airplane for the novice or the mediocre, and it was absolutely unforgiving close to the ramp". Designed as a bomber, it was never qualified as such, because its "linear" bomb-bay – consisting of an assembly of auxiliary tanks and a nuclear weapon, occupying a tunnel in the fuselage and ejected backwards through the tail – gave constant trouble up to the time that the US Navy carrier force renounced its nuclear strike role. Instead the Vigilante went into production and service as a reconnaissance aircraft. It was retired in 1981, mainly due to continued attrition of the force: even at that time, you could have parked a Vigilante alongside a number of aircraft just entering service, and an intelligent layman would have had trouble telling the veteran from the newcomer.

Sophisticated variable inlets, stability augmentation, electrically signalled controls: a great many features that become standard for later Mach 2 aircraft were pioneered on the Vigilante. Today's main Western fighters, the F-14, F-15 and Tornado, use the Vigilante intake design and echo its slab-sided, shoulder-winged styling.

By the end of the 1950s, therefore, it was clear that there were a great many other routes to Mach 2 besides the delta. More recently, however, the delta has experienced a revival as designers have shown increasing interest in "supercruise" configurations: combinations of aerodynamic and powerplant technology which will allow aircraft to sustain Mach 2 without using reheat, and thus to operate for extended periods at supersonic speed. Dassault has shown that some of the problems of the plain delta can be alleviated with modern control systems – descendants of the Vigilante system – based on high-speed digital computers, which react to aircraft movement with vastly superhuman speed and can control an "unflyable" configuration such as a flapped and slatted delta. Other designs are continuing the road embarked on by the Saab team three decades ago: the principle underlying the brand-new General Dynamics F-16E, with its thick, sharply swept inner wing rendered controllable by the thin, sharp-nosed outer section, is identical to the basis of the Draken's double-delta design. The F-16E's "cranked-arrow" wing, of course, emerged from a US effort to produce something better than the astonishing ogival wing of the Concorde supersonic airliner, in which powerful vortices shed from the wing roots create the same effect at low speeds as a blunt leading edge.

In November 1962 the British and French governments agreed to build the world's most advanced and ambitious Mach 2 aircraft; but, several months before, the prototype of an aircraft aimed at 50 per cent greater speed and similar efficiency had taken to the air for the first time, a milestone in one of the most difficult and successful aviation programmes of all time. In common with most of the world, the British and French ministers signing the November 1962 agreement knew nothing about it.

1400-2000mph

Standing the heat

1,400–2,000mph

Standing the heat

THE desolate country of high plains and dry lakes to the north of Los Angeles continued to echo to the sound of sonic booms long after Yeager's historic flight of October 1947. Muroc AFB – later renamed Edwards, after a pilot killed there in the crash of a YB-49 – was the centre of a vastly ambitious USAF project, aimed at pushing manned aircraft to speeds far beyond Mach 1.

As early as 1945 the USAF and NACA had contracted for the construction of three advanced research types. There was the XS-1, designed to be developed rapidly and to attain Mach 1 with minimum risk. There was the Douglas XS-3, intended for sustained flight at Mach 2. The fastest of all, though, was the Bell XS-2, a sweptwing aircraft more than twice as powerful as the XS-1, and intended to fly so fast that the heat created by air friction would cause aluminium to warp and fail. It would be constructed of materials such as stainless steel and K-Monel alloys, and was designed to reach Mach 3, or 1,980mph.

The XS-2 programme was dogged by problems from the start, both with the unconventional materials in the airframe and with its 15,000lb-thrust Curtiss-Wright XLR-25 rocket motor. Although metal was cut in late 1946, it was four years before an airframe was completed, and it took more than a year before it was ready for gliding tests. One reason for the delay was the continuation of the same manufacturer's XS-1 programme. The USAF had realised that the straight-winged aircraft had a great deal of speed potential, and in April 1948 it ordered four more X-1s – the "S" had been dropped from the X-series designations, reflecting the inclusion of experimental subsonic aircraft in the series. These were to have turbine-driven fuel pumps replacing the nitrogen system of the early aircraft, making room for considerably more fuel, and were expected to exceed Mach 2·4.

The first X-1 to be completed with a turbopump, however, was the X-1-3, third of the original batch. It never flew, having blown up on the ground in November 1951 during pre-flight checks. Of the longer-bodied second series of aircraft, the X-1D was the first to be completed, but on an early fuelled flight in August 1951 it suffered an explosion and had to be dropped from the carrier aircraft.

Meanwhile the more powerful X-2 was proving very dangerous to fly and to land. The first gliding flight ended in a ground-loop, but trials were continued in October 1952. Early in the following year the X-2 moved into the next stage: captive fuelled tests under its B-50 mother ship. In May 1953, however, the X-2 blew up as the X-1-3 and X-1D

had done, killing its pilot and one of the crew on the B-50 and nearly destroying the mother ship as well.

It was Chuck Yeager who made the first powered flight in a turbopump-equipped X-1, the X-1A, in November 1953. In typical Yeager style, he took the X-1A right to its performance limits within a month; on December 12 Yeager attained Mach 2·44, 1,612mph, at 74,200ft. At that point, in the thin upper air, the X-1A's control surfaces lost the battle with inertia coupling and the aircraft began to tumble end over end like a length of lead pipe. The tumble turned into an inverted spin, which became an upright spin, from which Yeager could recover by normal means – 51sec, 39,000ft and 1,440mph of airspeed later. As Yeager glided down to the dry lake he remarked to the controllers: "If I had a seat [meaning an ejection seat] you wouldn't still see me in this thing." The X-1A had one more battle with inertia coupling, following Maj Arthur Murray's flight to 90,000ft in the following year.

The X-1A's career ended in 1955, in the same violent manner as the X-1-3, X-1D and X-2, but mercifully without loss of life. Meanwhile, another X-2 – technically speaking, the first aircraft to be ordered – had been completed, tested and modified to improve its landing behaviour, while a satisfactory XLR-25 had been developed. After modifications, and a further interruption while the cause of the four X-series explosions was investigated (a secretion from the gaskets was to blame), the X-2 finally made a powered flight in late 1955, ten years after the programme had begun. In May 1956 Pete Everest took the X-2 to 1,670mph, or Mach 2·53, and two months later increased this speed to Mach 2·87. There was little doubt that the X-2 could attain Mach 3, but it was also clear that the aircraft's characteristics would be extremely unpleasant above Mach 2·7. Yeager and Murray had survived their encounters with inertia coupling; this may have encouraged the USAF to continue pushing the X-2 to higher speeds and altitudes. In September 1956 Capts Iven Kincheloe and Milburn Apt joined the X-2 project, and Kincheloe took the X-2 to 126,200ft, well beyond the point at which its controls were effective. On September 27 Apt dropped away from the B-50 mother ship on his first flight in the X-2, and went into an accelerating climb, quickly passing Everest's speed mark and hitting Mach 3·2, or 2,094mph at 65,500ft. Apt attempted to start a turn back towards the landing site, and the X-2 went into a vicious high-speed tumble; this time the movement was violent enough to knock the pilot unconscious. Apt did not recover sufficiently to escape and was killed.

NACA considered using the last of the second series of X-1s, the X-1E, as a substitute for the X-2 in a planned investigation of kinetic heating problems; the aircraft was modified with a high-energy fuel in an attempt to achieve Mach 3, but made only one flight before mechanical problems caused its retirement. The US companies which by that time

were working on Mach 3 service aircraft were left without a strong research base.

No fewer than three companies had started work on Mach 3 airframes in the late 1940s. All of them based their work not on the turbojet but on the ramjet, which was widely expected to replace it. The ramjet was an extension of the principle used by pre-war British aircraft to achieve zero-drag cooling; air entered a carefully shaped duct at high speed relative to the aircraft, was slowed down and simultaneously compressed, and was heated and expelled at higher speed. Instead of a radiator element the ramjet used fuel burners to heat the air. The principle had been patented in the early 1930s, by René Leduc of France, who had designed a ramjet-powered experimental aircraft before the war. In the USA Roy Marquardt had pioneered the development of ramjets, and in 1947 a P-80 was tested with a Marquardt ramjet on each wingtip. What attracted designers to the ramjet was that its pressure ratio, and thus its thrust and efficiency, increased rapidly with aircraft speed, and it promised to take over from the turbojet at high Mach numbers. On the other hand it had to be provided with some sort of boost before it would generate thrust, and its design presented magnified versions of the problems encountered by advanced turbojet inlet and exhaust systems.

Two of the Mach 3 airframes in question were unmanned "cruise missiles" in which the advantages of simplicity outweighed the weak points of the ramjet: they were the North American Navaho intercontinental strategic attack weapon and the Boeing Bomarc long-range interceptor missile. (The Navaho was cancelled in 1957, as soon as the fundamental problems involved in the more effective ballistic missile had been solved, while the Bomarc entered service in 1960.) The missiles could be launched by rocket boosters, eliminating the problem of providing a second powerplant. The third Mach 3 ramjet study was more advanced, stemming from studies of future manned interceptors by Republic in late 1947. The company decided that a combined turbojet/ramjet powerplant was best for a Mach 2+ interceptor, its designs becoming steadily more ambitious. Eventually, in September 1951, the US Air Force launched development of a series of new interceptors: the near-term Convair F-102A, designed to provide the USAF with a supersonic all-weather fighter as soon as possible; the developed Convair F-102B, with Mach 2 performance, and the Republic XF-103. The lat-

ter was by now a Mach 3 aircraft, powered by a Curtiss-Wright powerplant which combined a Wright J67 – a development of the British Bristol Olympus – with an XRJ55 ramjet. The powerplant was fed through a combined ventral intake which incorporated a complex set of variable ramps and flaps to allow operation in all-turbojet, all-ramjet and transition modes.

A full-scale mock-up of the XF-103 was completed in March 1953, and displayed many interesting features: the first Ferri-type intake with internal transonic compression, titanium structure and a novel configuration which featured a small 60° delta wing and a conventional tailplane; although the wing loading was high, the presence of the tailplane made it possible to fit effective flaps on the wing and thus keep low-speed performance within reasonable limits. The tailed-delta configuration was not widely used in the West, apart from the subsonic A-4, but distinguished a range of Soviet aircraft ranging from the diminutive MiG-21 day fighter to the monstrous Myashishchev M-50 strategic bomber.

Development of the XF-103 proceeded slowly, as might be expected for so radical an aircraft, although by late 1956 the designers managed to achieve transition from turbojet to ramjet power in the test-cell. The XF-103 was then aimed at the truly staggering speed of Mach 3·7, or 2,450mph, at 80,000ft. The team expected to complete the first aircraft in late 1958, spend a year in ground tests and fly early in 1960.

The XF-103 had taken so long to develop that it was now running almost in parallel with two other Mach 3 programmes started several years later. Both stemmed from North American, which at the time appeared to have monopolised the high-speed aircraft market in the USA, and both were powered by conventional jet engines. Oddly enough, both designs had appeared as the result of an official requirement for what at first sight seemed to be a much less difficult aircraft. This was Strategic Air Command's request for information on a successor to the B-52 long-range bomber, issued in late 1954 (on present plans, as these words are written, such a successor should enter service in 1986) and calling for similar range and payload, supersonic dash speed and the ability to operate from the same bases as the B-52. From this emerged two requirements: WS-125A for a nuclear-powered bomber, and WS-110A for a "chemically powered" aircraft. The term "chemical power" indicated that alternatives to conventional hydrocarbons were under consideration.

To begin with, the task appeared to be impossible because of the difficulty of reconciling the demands of efficient subsonic cruise with supersonic dash. Boeing and North American finally submitted designs based on a proposal by Dr Richard Vogt, the German designer who had originated the skew-wing configuration. Vogt's idea was to attach huge auxiliary wings and fuel tanks by flexible couplings to a high-speed delta aircraft, attaining a 260ft span without immense structural loads. Most of the weight would be carried by the outer wings, which would "float" without overstressing the couplings or the supersonic aircraft in the middle. Like so many of Vogt's ideas, it was extremely logical and worked as far as it was tested (in small scale on a light aircraft), but was utterly alien to the conventional sense of aerospace aesthetics. SAC's volatile chief, Curtis LeMay, rejected the idea outright.

Forced to rethink the concept, NAA engineers turned to a paper issued by Alfred Eggers and Clarence Syvertson of NACA's Langley research centre, which discussed a concept described as "favourable interference" or "compression lift". The idea was that patterns of high and low pressure caused by shock waves could be exploited to provide lift from energy that was otherwise wasted. If supersonic flight could be made more efficient, they reasoned, it might be possible to make the WS-110A cruise at supersonic speed all the way to the target.

Another influence was increasing confidence in the ability of turbojets to operate at very high Mach numbers. NAA and General Eleric had already proposed an advanced version of the J79 as the powerplant for a high-supersonic F-106 replacement: it was designated J79-X275, indicating its design operating Mach number of 2·75. The essential improvement that was promised to attain this sort of speed was efficient air cooling of the turbine blades; by blowing a thin film of relatively cool air over the blades, GE expected to be able to raise turbine-entry temperatures from 1,800°F to 2,500°F without encountering higher temperatures in the metal of the blade itself. This would allow the engine to operate at high Mach behind an efficient high-speed intake without exceeding its temperature limits.

A third influence on GE and NAA was the increasing interest being shown in exotic fuels. One in particular was a hydrocarbon derivative known as tetra-ethyl-borane, which incorporated highly reactive boron (in the same way that tetra-ethyl-lead, used in high-octane fuels, embodied a metallic element). Not only did boron fuels offer a great deal of extra energy per unit of weight, but the boron also acted as an oxidiser and allowed efficient combustion at higher altitudes. Boron-fuelled engines would give the same power at higher altitude than those running on conventional fuel; thinner air meant less drag, and less drag meant higher speed and longer range.

In March 1956 both Boeing and North American were put under contract to study supersonic-cruise bombers to WS-110A; meanwhile, another North American team was working on the WS-202 requirement for a Mach 3 interceptor. In the course of 1957, North American and GE won contracts to build the XB-70 Valkyrie bomber, the XF-108 Rapier interceptor and the J93-GE-5 engine which was to power both aircraft. The Republic XF-103,

Strategic cruise missiles are nothing new. The ramjet-powered North American SM-64 Navaho was one of the first air vehicles designed to cruise at Mach 3. Together with the supporting X-10 programme, the development of the Navaho broke a great deal of ground before the system was cancelled in 1957

which was designed to do the same job as the more flexible F-108, was cancelled in August of that year.

The key to victory on the XB-70 was aerodynamics: North American promised a lift/drag ratio that was 15–20 per cent better than its rival considered possible, almost entirely due to compression lift. Aerodynamically, the Valkyrie was extraordinary, even though it was based on a plain 65° delta. To begin with, it was a tail-first design with a large thin-section foreplane mounted on the forward fuselage. Although this added weight to the structure, it provided upwards trimming force at low speeds – when the delta wing has to operate at high angles of attack – and at high Mach numbers, when the aerodynamic centre of the wing tends to shift backwards and nose-up trim is needed to maintain level flight. Thus the canard reduced the need for trim downloads and made it possible to make the wing smaller.

One feature of the Valkyrie remains unique to this day: the outer wing panels folded downwards through 65° – more than half-way to the vertical – at high speed. This monstrous exercise in variable geometry added directional stability, and alleviated the rearward shift of aerodynamic centre with increasing speed by simply reducing the wing area

at the back. Additionally, beneath the delta wing was a huge Gothic-arch-planform body which contained the six J93 engines in line abreast – fed through two gaping intakes – the Valkyrie's weapon bay and the landing gear. Both this wedge-body and the downturned tips would generate powerful shockwaves, which would then be trapped between the tips and the centrebody. The idea was that the whole thing would work like a gigantic supersonic inlet, supporting the 200-ton weight of the Valkyrie at 70,000ft or higher like a surfer riding a wave.

North American chose stainless steel as the main structural material for the XB-70. Most of the aircraft was to be made of stainless-steel honeycomb – two skins of steel brazed to a core made of hexagonal segments, so that a cross-section resembled an open honeycomb. Parts of the airframe were expected to attain tremendously high temperatures, so the fuel was pre-cooled and used to cool various systems before being fed to the engines. Special techniques had to be devised to cool the wheels and prevent the tyres burning out during high-speed flight.

The original schedule called for the first flight of the XB-70 in mid-1963, but soon after the contract was signed it was decided to accelerate the programme and fly by about the end of 1961. The

Rapier, which shared a great deal of technology, and the engine and boron-fuel programme costs, with the XB-70, was running in parallel. But technical troubles were as nothing to the political and financial problems. The Soviet Union had orbited the first artificial satellite in October 1957, and everybody was talking about a "missile gap". The Atlas, Titan and Minuteman intercontinental missiles went into production in quick succession, calling for the development of new technology and the construction of complex new bases. The US Air Force was caught in a budgetary trap, and was ready to listen to those who were saying that manned fighters and bombers were obsolete, due for replacement by the missile. In the second half of 1959 the high-Mach programme was eviscerated. The entire boron-fuel project was cancelled in August, after persistent problems: the fuel clogged anything that would clog, and was toxic, so the J93-GE-3 with conventional fuel was substituted in the B-70. The next month the F-108 Rapier programme was cancelled outright. (The US Air Force had undisclosed reasons for doing this, as we shall see later.) The final cut came at the end of the year, when the B-70 programme was reduced to a single prototype.

For a few months, starting in mid-1960, the B-70 was granted a temporary reprieve; but the incoming Kennedy administration was disposed to regard the supporters of the bomber as romantic veterans. In March 1961 plans to order 12 operational B-70As were cancelled and the programme was once more cut back, to three prototypes this time. (The third was cancelled later.) Congress voted money to reinstate the programme in July, but Kennedy vetoed the proposal. The state of the high-Mach programme seemed desperate; those who knew otherwise were few, and they made no effort to enlighten those who were not cleared to know better. The better informed group included those in the USAF who had finally decided to cancel the F-108: why develop an aircraft yourself, when another organisation has decided to pay for a similar job on its own budget? Just before the F-108 was scrapped, the decision was taken to build a highly advanced Mach 3 aircraft, rather bigger than the Rapier, using funds which were beyond the reach of the missile planners and administered by an agency which the pro-missile "whiz-kids" with their computer studies could not touch: the Central Intelligence Agency.

One of America's best-kept secrets in the 1950s was the fact that US aircraft were systematically violating the sovereign airspace of the Soviet Union. Their mission was strategic reconnaissance: gathering information on the military technology and industry of the Soviet Union, most of which was based in areas completely closed to all foreigners. Up to 1960 the US Air Force had managed to use aircraft with sufficient height capability to penetrate the thinly spread Soviet interceptor screen. Strategic Air Command was the responsible divi-

sion, and used stripped-down B-29s and even higher-flying B-36s, soaring to altitudes of 55,000ft and more. In 1954 Kelly Johnson of Lockheed made a new proposal directly to the CIA: that Lockheed should develop a special reconnaissance aircraft for the agency, capable of sustaining greater altitudes than any foreseeable fighter and operating outside the range of first-generation missiles. This became the Lockheed U-2, which flew in August 1955 and made its first flights over the Soviet Union in the following year. Even then it was clear that Soviet defensive weapons would eventually catch up with the U-2. Lockheed studied ways of improving the U-2 – making it fly higher, making it less visible to radar, or adding more effective jamming equipment – but the aircraft was very sensitive to increased equipment weight and even the installation of a more powerful engine did little more than offset the extra weight of added reconnaissance and defensive gear. Lockheed's thoughts turned by 1957 towards a completely new aircraft, much faster than the U-2 as well as higher-flying, with two engines and a heavier sensor load.

Lockheed did a great deal of work on one project that would have used liquid hydrogen (LH2) as its fuel; although LH2 is much less dense than a hydrocarbon, and has to be carried in insulated tanks, it can release far more energy per unit of weight and thus makes for a very efficient aircraft. However, the complications of supplying an operational aircraft with a cryogenic fuel caused Lockheed to drop the project.

From early 1958 Lockheed and a number of other companies made a variety of presentations to the CIA. Lockheed settled on a turbojet-powered, hydrocarbon-fuel concept; General Dynamics (Convair) proposed a Mach 4, ramjet-powered aircraft launched from a modified B-58. At the end of August 1959 Lockheed's design, known within the company as the A-11, was declared the winner, and in January of the following year, a few months before one of the CIA's U-2s was shot down over Sverdlovsk, the CIA ordered 12 of the new aircraft. Apart from a single report, in the wake of the U-2 incident, that the CIA had already ordered development of a supersonic replacement for the type, not a word leaked out about the project until President Johnson revealed its existence at the end of February 1964. By that time, the aircraft had been flying from a closely guarded base in Nevada for two years.

Although it was hoped that the A-11 programme would benefit from technology developed by the X-15 rocket research aircraft and the XB-70 bomber, this never came about. The X-15's short Mach 3 flights provided no data on the most serious design problems: the aerodynamic difficulties of developing a stable but efficient Mach 3 configuration, the design of a Mach 3 air-breathing powerplant and how to achieve sustained flight at high temperatures. The XB-70 provided little data for

In sharp contrast to the Valkyrie was Kelly Johnson's sinuous A-11. Johnson did not believe in compression lift, which would in any case have conflicted with the need for "stealth" design. The resulting aircraft resembled a Mach 3 surfboard, as befitted a Southern California company *(Lockheed)*

two reasons: most of the solutions chosen by Kelly Johnson and his team were different from those selected by North American, and the A-11 programme rapidly overhauled the XB-70 project.

Lockheed rejected both canards and compression lift in the design of the A-11, which was complicated by the requirement that the aircraft should produce the smallest possible image on a radar screen; even had Lockheed believed in the claims made for compression lift, the sharply angled and slab-sided bodies needed to exploit it would have given the aircraft a very large radar cross-section (RCS). Instead the aircraft emerged as a basically conventional delta; but to minimise the side-view profile of the aircraft, an important element of RCS reduction, the engines were placed midway along the wings. The fuselage was basically very slender, but was smoothly blended into broad horizontal chines which ran right forward to the nose, while the entire forward fuselage was bent slightly upwards compared to the wings. The chines fulfilled a triple purpose. First, they acted as a very low-aspect-ratio foreplane, generating lift at high Mach, trimming the nose upwards and reducing the need for trim downloads at the rear. Secondly, they "streamlined" the fuselage against aerodynamic side forces and allowed the A-11 to be controlled at low speeds by two small, low-RCS fins (each comprising a fixed stub and an all-moving upper section). Thirdly, they created a flat blended shape which further

reduced RCS; the engines were similarly blended into the wing. The all-moving fins – rudders had been ruled out because of the high temperatures developed in the stagnation area around the hinges – were canted slightly inwards, reducing rolling moment due to sideslip and incidentally reducing RCS.

The immense subtlety of the A-11's aerodynamic design was not appreciated for a long time, mainly because the importance of low RCS or "stealth" was a very well kept secret until the mid-1970s. Lift/drag ratio at Mach 3 is about 6·5:1, and this is the main contribution to the long range of the A-11 and its descendants. The A-11 broke new ground in another respect: at high speeds it is unstable in pitch and yaw. Because of this, it was not only necessary to include a stability augmentation system (SAS), but it also had to be as reliable as a conventional, "natural" system because the aircraft would be unflyable without it. This was provided by a three-axis, eight-channel AFCS developed by Honeywell. Lockheed considered the installation of an electrically signalled (fly-by-wire) control system, but decided that there were too many unknowns involved.

The engine chosen for the A-11 had already run some 700hr before the Lockheed design studies started. It had been developed by Pratt & Whitney as the J58-P-2, intended for use on a Navy attack aircraft with a Mach 3 dash capability. It was, in its original form, typical of a 1950s "all-supersonic" engine, designed with a moderate pressure ratio (it was a single-shaft engine without extensive variable geometry) so that it could function downstream of a highly efficient inlet without exceeding design temperatures. As P&W and Lockheed started to work on the A-11 powerplant, however, they discovered that at high Mach numbers the engine could not swallow as much air as the inlet would deliver; the result was a loss in efficiency and thrust at high speed. P&W accordingly modified the J58 to incorporate an annular fixed bleed valve behind the fourth stage of the compressor, feeding excess air into six large pipes along the sides of the engine. These delivered fourth-stage air directly to the afterburner, cooling the airflow to the burners and allowing the reheat to develop a greater temperature increase – and hence a greater thrust increase – without exceeding airframe temperature limits.

The engine was fed through an axisymmetrical mixed-compression intake, consisting of a circular inlet blocked by a huge conical spike. The spike could be moved 26in fore and aft by powerful hydraulic actuators (they had to be powerful, because loads on the spike can exceed 30,000lb). At rest the spike is in the forward position and the inlet throat is as wide as possible. In addition the J58 breathes through a series of doors and ducts in the nacelle, which feed the inlet through two sets of valves: one set in the spike itself and one in the cowl farther aft. As the aircraft accelerates, these valves first close. At transonic speeds shock waves form between the spike and the inlet wall; the intake is then said to be "started", and the spike gradually retracts under automatic control: the position of the spike is basically determined by Mach number, but angle of attack, sideslip and acceleration are also taken into account. As this happens, the cowl bleed which helped to feed the intake at low speed reverses, extracting air from the inlet; this "traps" the shock wave by creating a pressure reduction behind it, and at the same time supplies cooling air for the engine. Finally, there is another bypass door at the rear of the inlet, which opens as necessary to dump excess air down the cowling to the exhaust nozzle. The latter was the first of the "ejector" type: it comprises a large ring, forming part of the airframe, with annular doors in front and a variable-area nozzle at the back. At low speed and thrust levels, the ejector draws in outside air through the front doors to reduce drag and increase the propulsive efficiency of the engine. At high speed the engine and afterburner provide enough airflow to fill the ejector area, so the doors at the front close (at about Mach 1·1) and the flaps at the rear of the ejector open to provide a wide, efficient propelling nozzle.

The net result of this uniquely complex system is that at high speed and altitude the J58 is a cross between a turbojet and a ramjet. The intakes can produce a pressure ratio of 40:1 at maximum speed, and compression in the inlet produces 60 per cent of the engine's thrust. A good percentage of the inlet airflow bypasses the combustors and turbine stages, and feeds either into the afterburner or the ejector nozzles. The air leaves the afterburner at roughly the same speed as it entered the compressor, but hotter and more compressed; it mixes with fresh, inlet-compressed air in the ejector and is re-accelerated to produce thrust.

The passage of the A-11 through the atmosphere at Mach 3 generates enormous amounts of heat, and like other companies working in this area Lockheed realised that aluminium would be virtually useless. The only usable materials were stainless steel or titanium; while the latter was stronger, the technology for building a whole airframe from titanium simply did not exist, whereas the properties of stainless steel honeycombs were comparatively well known. Despite this Johnson elected to build the A-11 from titanium, using conventional structural design. Tooling and quality control for titanium – which is strong, but tricky to use and highly reactive – proved to be an enormous task. At the outset the drills used to make fastener holes would last for only 17 holes; Lockheed eventually pushed drill life to 100 holes, using new cutter materials, lubricants and techniques. A quality control programme was developed under which every individual part (13 million-plus throughout the programme) could be traced back to the point where the chunk of metal it came from was poured in the mill. Another problem was that fasteners would simply fail if they were adjusted with cadmium-plated tools; every wrench in the entire programme had to be checked to weed out such tools.

Even for pure titanium, the A-11 was a hot environment. Not a single part of the external surface was below oven temperatures; the bulk of the airframe "cooked" at between 500° and 600°F; the engine nacelles, heated from outside by air friction and from the inside by the working cycle of the engines, attained between 950° and 1,050°F over most of their length, while the tailpipes and ejectors were, quite literally, *white hot* at nearly 1,200°F. This posed some interesting questions of structural design. The wing skins were found to warp in early laboratory tests, so Johnson took a lesson from Hugo Junkers and corrugated the wing skins. Then there was the long forward fuselage, and its effect on trim. Its weight and hence its tendency to bend down varied with the fuel load, but there was another complicating factor: as the fuel level in the tanks fell, the fuel cooled the lower surface of the fuselage, but not the upper skin. The consequent differential expansion made the fuselage bend downwards. Thermal expansion also made it very difficult to seal the fuel tanks, so the problem was simply avoided: the tanks are sealed as well as possible, but leak heavily while the aircraft is cold. When it is hot, all the joints and fasteners snug up and the tanks become leakproof again. Parts of the A-11 secondary structure are made from a still-secret plastic honeycomb developed by Lockheed.

Unlike the XB-70, which used JP-6 fuel impregnated with dry nitrogen before fuelling, the A-11 had to offer basing flexibility and the potential for refuelling in flight to meet the CIA's need for global coverage with a small fleet of aircraft, independent of large facilities overseas. For this reason the A-11's fuel had to be free from requirements for special handling. Shell developed a basically hydrocarbon fuel called JP-7, meeting the extreme temperature requirements: from $-90°F$ in subsonic cruise to well over $350°F$ at high speed. (This was why the leaky tanks could be tolerated. An aircraft that sits in a puddle of fuel would usually be a safety officer's nightmare, but with JP-7 it is reasonably safe.) However, the use of unrefrigerated fuel meant that the fuel had capacity to cool the crew compartment, the nitrogen-filled tyres and the mission electronics, and that was all. Everything else simply had to survive the ambient temperature, whatever that might be.

In the engine, a Pratt & Whitney engineer later noted, this meant that "there was not one part, down to the last cotter key, that could be made from the same materials used on previous engines". The whole engine – discs, shafts, casings and all – had to be made from the sort of materials previously used for turbine blades. All valves and nozzles were moved by a one-pass system using fuel for hydraulic fluid; the fuel was burned immediately. There was virtually no cooling capacity for electronics, so there were only two electronic or electrical devices on the engine. The only way to achieve stable exhaust gas temperatures was by a manually operated control in the cockpit, working a small electric motor buried in the hydromechanical fuel control; a conventional automatic EGT trimmer would not work at the high ambient temperatures. The accessory gearbox was located as far as possible from the heat of the engine, connected to it by a long shaft. The ignition system was unique: conventional electrical ignition would be inadequate to fire up the high-flashpoint JP-7, even if it could survive the temperatures within the engine, so the designers switched to a chemical ignition system using tetra-ethyl-borane (TEB), the "zip fuel" originally intended for the B-70, to ignite the combustors and the afterburners. It is not known whether TEB has ever been used to boost the performance of the aircraft at high altitudes.

Almost incredibly, Johnson's team had the first A-11 complete some 30 months after the design was declared the winner. Pratt & Whitney's advanced project group at West Palm Beach, Florida, had taken a little longer – not one component of the A-11's JT11D-20 was common to the original J58-P-2 – so it was decided to check out the low-speed handling using a pair of J75s. One winter's night the first A-11 was trucked out of building 309 at Burbank and hauled across the mountains to "the ranch" – the secret base in the middle of a huge Air Force range in Nevada where the U-2 had made its

first flight. The A-11 took to the air on April 26, 1962. A few months later the J58 passed its preflight tests and one of the new engines replaced one of the J75s; after preliminary checks, two J58s were installed.

The first engine problem to be encountered was that the J58 would not start. Starved of air, the intake simply sucked what it needed out of the engine, via the fourth-stage permanent bleed. To begin with Lockheed removed an access panel for starting, and then added another set of suck-in doors.

The airframe presented remarkably few problems: the ejector had to be redesigned due to excessive drag, and the aircraft burned too much fuel going transonic. The latter problem was never completely solved, but is simply avoided in service by going transonic in a shallow dive. Such problems paled into insignificance beside a new, alarming and apparently random occurrence: inlet unstart.

"Unstart" was to the supersonic inlet what stall was to the turbojet. Somewhere downstream the inlet would choke, and the build-up of pressure would throw the primary shockwave out of the inlet. The engine would lose power, drag would increase violently; the aircraft would yaw viciously into the dead engine and simultaneously roll. The AFCS was programmed to slam 9° of rudder (quite a lot at Mach 3) in one-seventh of a second, and this would cause the aircraft to lunge back almost as violently. One problem was that the violence of the programmed reaction often confused the pilot (whose head was still bouncing off the canopy) as to which inlet had unstarted.

Unstart was never entirely cured, but was reduced by the introduction of extra bypass doors (the aftermost pair, which are closed to increase thrust at high Mach), an automatic trimming system (with a manual back-up) and better sealing of the nacelle. In early development the pneumatic control system for the inlet spike had to be replaced – despite heat and vibration – by a more accurate electronic system.

By early 1964 the Skunk Works had built and flown eight of the 12 aircraft originally ordered by the CIA, and the aircraft had been used for operational overflights. The US Government wanted no repetition of the U-2 affair, and had no problems in that respect; cruising at Mach 3 and up to 100,000ft, the A-11 was out of range of almost any defensive system in the world. The one exception had been ordered into development in mid-1960 and flown in August 1963; this was the YF-12A, which combined the radar and missile system originally developed for the F-108 with a modified A-11 airframe. The radar required the installation of a conical nosecone, breaking the smooth lines of the chines and disrupting airflow so that three ventral fins had to be added. (The later YF-12C had a modified radome which overcame this problem.)

Unlike the U-2 the A-11 could be refuelled in

flight; its operations could therefore be confined to the secret Nevada airfield, which was surrounded for a great distance with signs advertising the presence of unexploded munitions. This tended to deter the curious, so President Johnson's announcement of the existence of the A-11, on February 29, 1964, came as a complete surprise. Defence Secretary Robert McNamara's assertion that "the A-11 is an interceptor, it is being developed as such, and beyond that I have nothing to add" fooled nobody. The origins of the A-11, the concealment from budget scrutiny of hundreds of millions of dollars in development funding, plus the known complete lack of any Russian bomber threat that would justify such an interceptor, all pointed to the true role of the aircraft.

Kelly Johnson, however, had more ambitious goals for the A-11 design. In early 1961 Lockheed made the first proposals to the USAF for a bigger and heavier version of the A-11, designed for strategic nuclear strike as well as for reconnaissance. About that time SAC officers were making a last attempt to save the Valkyrie programme by abandoning the "dated" bomber label. Instead they were promoting the accuracy and flexibility of the manned aircraft, and its ability to see its targets before committing to an attack, and they redesignated the North American aircraft as the "reconnaissance/strike" RS-70. Lockheed felt that a version of the A-11 would accomplish the same mission, but ran into opposition not only from the anti-bomber forces within the USAF, but also from the RS-70 lobby. Nevertheless the company was encouraged to build a full-scale mock-up of the reconnaissance/strike aircraft, and this was complete by mid-1962. At the end of the year, all reasonable hopes for the reinstatement of the Valkyrie programme having withered, SAC ordered the first batch of improved aircraft under the designation Lockheed RS-71.

This development was announced by President Johnson in July 1964. It was described as a new type – not strictly accurate – and the designation was given as SR-71, not RS-71. Lyndon Johnson himself had inadvertently transposed the letters at his press conference and a correction would have proved embarrassing during an election campaign. Therefore it was as the SR-71 that the aircraft flew in December 1964, entering service at the beginning of 1966 with the 9th Strategic Reconnaissance Wing; this unit had taken over the role of the by then thoroughly "blown" covert CIA units. Lockheed's efforts to secure a large order for armed SR-71s, as replacements for some of SAC's older B-52s, had been dealt a final blow in late 1965 when the cheaper General Dynamics FB-111A was chosen as an "interim" strategic bomber pending development of the Advanced Manned Strategic Aircraft (AMSA). About 30 SR-71s were built, including at least two trainers – the SR-71B and C – with raised rear cockpits, and all were delivered to the 9th SRW at Beale AFB, California.

The SR-71 uses the same powerplant as the A-11, and has a longer fuselage with a refined chine shape. It also has a very much larger internal fuel capacity and a maximum take-off weight of 170,000lb, some 35,000lb heavier than the A-11/YF-12. About 25,000lb of this increase is believed to be fuel. The SR-71 does not usually take off at maximum weight, because its performance would be marginal if one engine failed, but instead takes off with a reduced fuel load and makes an immediate rendezvous with a KC-135Q tanker.

The SR-71s of the 9th hold world records for speed, at 2,189mph, and altitude in sustained flight, at 86,000ft. With tanker support and several Mach 3 excursions, one SR-71 accomplished a 15,000-mile flight in 10·5hr in April 1972, demonstrating the global range of the aircraft; SR-71s have been used to reconnoitre some of the world's most intensive conflicts, and have overflown the Middle East and North Vietnam. By the late 1970s, more than 800 surface-to-air missiles had been fired at SR-71s, and not one had been brought down. Together with virtually all the unclassified evidence, this suggests that the operating altitude quoted for the SR-71 by the USAF – "above 80,000 feet" – is a few miles on the conservative side. It has been reported that the original CIA requirement called for a 20-mile (about 106,000ft) operating altitude, and it is very possible that the SR-71 can sustain heights not far short of that.

The SR-71 has been semi-officially dubbed Blackbird because of its sooty blue-black finish; this is claimed to release more heat by radiation than the airframe gains by the extra friction of a matt rather than glossy surface, particularly in the thin air of high altitude. However, the similarity of the colour to that of the subsonic U-2R lends weight to the suggestion that it is a form of radar-absorbent finish.

Today, twenty years after the first flight of the A-11, Kelly Johnson's design remains by a useful margin the fastest and highest-flying aircraft in the world. Two other Mach 3 service-type aircraft have been flown since the A-11. The first to appear was the XB-70, which made its first flight from NAA's base at Palmdale to Edwards AFB in September 1964, just before the YF-12 was presented to the world's press, and two years behind schedule. The most serious delays had involved difficulties in sealing the fuel tanks, the problem which Lockheed had side-stepped. In October 1965 the XB-70 attained Mach 3 for the first time, but neither of the two prototypes ever demonstrated anything like their design range of 7,000 miles at Mach 3. The second XB-70 was wrecked in a collision in June 1966, while leading a formation of GE-powered aircraft for a publicity film, and the first was finally retired in early 1969.

The other Mach 3 aircraft to have been revealed so far appeared as a direct consequence of the US Air Force and CIA efforts, but was utterly different

in concept. It was developed to meet a requirement formulated in 1959–60 by the Soviet Union's Air Defence Forces (PVO) for weapons to defend the country against Mach 3, high-altitude intruders such as the B-70 and the advanced reconnaissance aircraft planned by the CIA. This was not as easy a task as some of the detractors of the manned bomber implied. The SA-2 Guideline, the missile which had downed the U-2 in May 1960, was too slow and limited in ceiling to catch an almost equally fast adversary, forewarned by its ECM. Only a very large missile, tied to fixed sites, or a manned interceptor of extremely high performance offered a reasonable chance of success, and development of both was pushed ahead. (The missile which resulted is the large weapon known to NATO as the SA-5 Gammon.)

Both the Mikoyan and Sukhoi bureaux had, by 1960, scaled up their tailed-delta Mach 2 fighters into larger prototypes (the Sukhoi T-37 and

Mikoyan I-75), with large axisymmetrical mixed-compression inlets and some steel in their construction, which could attain speeds between Mach 2·5 and Mach 2·8. The interception mission called for a new approach, however, because it could be assumed that the target would carry heavy ECM. The aircraft therefore had to have a powerful radar, combined with a battery of infra-red and radar-homing missiles; all of these conspired to make the interceptor a great deal larger and heavier than previous Soviet target-defence fighters, until it approached the weight of the long-range Tu-28P.

The PVO selected a design from the Mikoyan bureau to meet the requirement. Mikoyan's inspiration had clearly come from North American – but not from the XB-70 or F-108. The E-266 prototype, as it appeared when rolled out in the course of 1964, was so like the A3J Vigilante that somebody should have sued. The massive two-dimensional inlets with internal ramps and dump doors; the slab-sided

fuselage with shoulder-mounted wings and mid-set tailplane; the large-area, moderately swept, thin-section wings; the single-wheel main gears retracting forwards and upwards into the fuselage sides: obviously the Mikoyan bureau had seen the right aircraft to copy. The E-266 had twin fins, slightly angled outwards, in contrast to the single vertical slab on the Vigilante; but the fins could have come straight off the original Vigilante mock-up, before North American changed the design. North American, later Rockwell, continued to try to push the Vigilante as an improved interceptor for the US Air Force right up to the early 1970s, but without success.

There were some detail differences in the aerodynamic design. Unconstrained by the tight take-off performance requirements which governed the design of the American aircraft, the Mikoyan team could use a smaller wing with simple flaps and conventional ailerons, while the fins and slab tails were cut off obliquely in the classic Soviet way of solving flutter problems.

Structurally, the E-266 was much more interesting. The airframe was not built of exotic titanium alloys, nor was it constructed of intractable stainless steel. The aircraft needed only to withstand short exposure to high temperatures, so the Mikoyan bureau elected to hand-weld it out of mild steel: tough, resilient and relatively heatproof. Arc-welding avoided many problems of thermal expansion. Mikoyan's solution to the problem of sealing nitrogen-blanketed fuel tanks was as direct as Kelly Johnson's: the Soviet designers simply abandoned the idea of integral tanks and filled every available part of the airframe with continuously welded steel fuel cells. They could expand and contract to their

hearts' content and never leak a drop.

The engines of the E-266 were another example of its superb suitability for its intended role. They stemmed from a Tumansky design originally intended for once-only use on high-speed drones or cruise missiles. The design carried to extremes the 1950s concept of an "all-supersonic engine" designed to operate behind a high-pressure inlet system: very simple, with a low static pressure ratio, but designed to operate at high temperatures. The Tumansky R-31 (sometimes referred to as the R-266) has only five compressor stages and a single-stage uncooled turbine, and has a static pressure ratio of 7:1. This makes it hopelessly inefficient at low speeds, but the overall pressure ratio of the powerplant, including the electronically controlled inlets, steadily increases with high Mach number. At high speed, a mixture of methanol and water is injected into the intake to cool the incoming air and hence the engine; for a short period this smooths out the mismatch between intake and engine airflow which P&W discovered on the original J58 and cured with the bleed-bypass cycle.

The E-266 was revealed to the world at the great Domodedovo air show in 1967, and shortly afterwards set a startling burst of closed-circuit speed records at Mach 2·8. Among the records broken were several set by the YF-12 two years before. It went into service with the Soviet Air Forces in 1971–72, with the designation MiG-25, as an interceptor (although the SR-71 was the only aircraft that could pose a real interception challenge) and a tactical reconnaissance aircraft. In 1973 a MiG-25R reconnaissance aircraft operating from Cairo was tracked at Mach 3·2 over Israel (it landed with its engines burned out), and in the same year an E-266

129

development aircraft set a string of time-to-height records. The USA's F-15 beat them in early 1975, and the MiG-25 took them back a few months later with contemptuous ease, carrying two tons of ballast to boot.

By this time the US intelligence community was desperate to find out how the MiG-25 could be so fast without looking like a B-70 or an SR-71. In September 1976 their prayers were answered: at the airport at Hakodate, Japan, an All Nippon Airways Boeing 727 was told to clear the runway immediately, because an unidentified aircraft was approaching without clearance and at high speed. It was a MiG-25 interceptor, flown by a Soviet Air Force officer who had become disillusioned by the system and (possibly) had some idea of just how much the

West would like to see his aircraft. Intelligence experts stripped the aircraft down, and after a few months of culture shock realised that the MiG-25, like any aircraft, had its strong and its weak points: high speed had been obtained at the expense of combat radius and flexibility.

In 1982 it appears that the Mikoyan bureau – now led by Rostislav A. Belyakov – may have addressed and rectified some of the MiG-25's failings in the shape of the new Foxhound, or Super MiG-25, with a more effective missile system and better range and endurance. Meanwhile, however, not even the MiG-25 has succeeded in catching the swift, stealthy and high-flying king of all air-breathing vehicles: Kelly Johnson's elusive and mysterious Blackbird.

Aggressively angular, the MiG-25 Foxbat is in complete contrast to the Lockheed series and is in many ways a remarkable design achievement. The MiG-25 has now been developed into the formidable Foxhound interceptor (*Motor-Press International*)

2000+mph

To space and back

2,000+mph

To space and back

IN October 1942 the bleak Baltic peninsula of Peenemünde shook to a new sound: the roar of a 55,000lb-thrust engine, driving an A-4 rocket on its first successful flight test. A critical milestone in the most extraordinary and most innovative aerospace programme of its day had been passed; the feasibility of all the new features incorporated in the A-4 had been proved. The automatic control system worked; so did the little vanes in the exhaust stream, which vectored the engine thrust to control the missile at speeds where the aerodynamic controls were ineffective. Liquid oxygen, only stable at subzero temperatures, had been proved to be a practical fuel. The turbine-driven pumps feeding 280lb of fuel every second to the combustion chamber had functioned properly. The engine, 15 times more powerful than any other aerospace powerplant even conceived at that time, also performed as it was supposed to. The A-4 programme may have been a strategic mistake – it has been remarked that the damage it did to the Luftwaffe through its voracious demands for scarce human and material resources could hardly have been greater if it had been designed for the job – but its technical achievement was staggering.

As compatriots and colleagues of the German rocket pioneer Hermann Oberth, the team working on the A-4 project were well aware of the physical theories which deemed it possible to orbit an artificial satellite around the earth, and prepared design studies for multi-stage long-range missiles and orbital probes. Alongside these projects, however, German researchers examined a different concept: the hybrid ballistic/aerodynamic aerospace vehicle.

The basic motivation for these studies was that the A-4 missile in its ballistic trajectory struck the ground at 2,000mph: this was far more than was necessary to evade the defences of the day, and represented wasted energy. The design head of the A-4 programme, Wernher von Braun, decided to recover some of that energy by fitting wings to an A-4, so that it could descend in a high-speed glide rather than a ballistic plunge. The range of the weapon could be more than doubled, von Braun calculated. The resulting A-9 was considered both as a replacement for the A-4 and as part of a two-stage weapon intended to attack the eastern USA. In the latter role it would be launched by an A-10 booster and would attain a maximum speed of 6,300mph before starting its 2,500-mile dive towards the target.

Peenemünde was the only place in the world at that time where such a design could be contemplated. By the end of 1940 von Braun's team had commissioned a wind-tunnel capable of testing objects at Mach 4·4, very nearly the maximum velocity of the A-4, and this facility was used for refining the shape of the A-9 as early as 1943. The final configuration had small, thin, swept-back wings of very low aspect ratio. Design work was suspended in late 1943 because A-4 problems were demanding attention, but was resumed a year later because the need for greater range was urgent: invading forces were pushing A-4 launching sites out of range of the weapon's intended targets. In early 1945 the A-4b winged test vehicle reached 2,700mph, by far the highest velocity attained by a winged machine.

Another project on the Peenemünde drawing boards was a manned aircraft based on the A-9, with a tricycle landing gear, flaps and airbrakes. Like the A-9 it would have taken off vertically, reaching an altitude of 60–65,000ft and a speed of 2,800mph before cutting the motor and coasting up to 95,000ft. The pilot would then guide it down to a conventional landing. The duration of the flight would be about 17min and a total distance of 400 miles would be covered.

Advanced and ambitious as that project may have been, it paled beside the proposal by Eugen Sanger and Irene Bredt for a long-range rocket-powered military aerospace craft, weighing 100 tons and launched at 1,000mph from a colossal rocket sled. The idea was that the craft would be boosted beyond the atmosphere, but not into orbit; instead it would "skip" across the upper surface of the atmosphere as a stone skips across a pond. Between them the piloted A-9 and the Sanger-Bredt "antipodal bomber" were to inspire a great deal of post-war work.

The bulk of development effort in the late 1940s, however, was directed at the more mundane problems of supersonic flight, rather than the speeds dubbed "hypersonic" beyond Mach 4. Until the early 1950s, the piloted A-9 and the Sanger-Bredt design remained curiosities with few practical uses. But the steadily increasing speeds foreseen for production aircraft – the Mach 3 XB-64 Navaho and XF-103 were under way by 1951 – suggested that the next wave in development for aggressive and defensive aircraft might well be beyond Mach 4 and above 100,000ft. At that point, though, the term "aircraft" becomes inadequate. Although the vehicle would have to have the characteristics of an aircraft it would operate at heights where there was not enough air to sustain combustion of fuel or to make conventional control surfaces effective. On the other hand there would still be enough air to make its presence felt, and in no uncertain fashion. Around the front edges of the vehicle a layer of air would be trapped by the speed of the object. As speeds increased the trapped layer would become steadily more compressed and would reach terrifying "stagnation" temperatures. While the problem with Mach 3 aircraft was to make their structures durable and efficient, the problem at higher speeds was to prevent the stagnation pressures and temperatures from ripping the airframe into shreds.

Both NACA and the US services began to study the problems of hypersonic flight around 1952–53. NACA had by that time perceived the limitations of the Bell X-2, and was looking at the qualities needed for a research aircraft capable of Mach 10 and a 50-mile altitude – out to the point where the atmosphere was effectively non-existent. The Office of Naval Research, which had been responsible for the Skystreak and Skyrocket programmes, was pursuing similar objectives and had issued a design contract to Douglas for a hypersonic aircraft in the same series, the D-558-3. Meanwhile one of the architects of the V-2 programme, Walter Dornberger, was working for Bell and raising considerable interest in boost–glide concepts; the USAF was undertaking a study labelled MX-2145 of the types of bomber it might need in 1965–70.

Douglas made considerable progress with the D-558-3 design, arriving at a 22,000lb aircraft powered by a 51,000lb-thrust liquid oxygen/ammonia rocket from Reaction Motors and spanning only 18ft. It would be covered by a thick magnesium skin

to absorb the heat generated by its passage through the air, and the leading edges of the wings would be solid copper; the idea was to conduct heat as quickly as possible away from the hottest parts of the aircraft. The extremities of the aircraft carried tiny hydrogen peroxide rockets to stabilise and control the aircraft outside the atmosphere.

But the D-558-3 was overtaken by a decision to merge the Navy programme with NACA and Air Force efforts, which led to the issue of a requirement for a hypersonic rocket aircraft in late 1954. The aircraft was to attain over 4,100mph: twice as fast as the expected maximum speed of the X-2, which had yet to fly under power. A design from North American beat the D-558-3, as well as other submissions from Bell and Republic, and was ordered as the X-15. Construction of three aircraft started in September 1956.

The X-15 resembled the Starfighter or the X-3, with thin-section trapezoid wings. The vertical tail was a thick, sharp-cornered wedge shape; the body a tank for nearly 2,500 US gal of LOX and ammonia. The airframe was welded together from Inconel-X, one of the chrome-nickel alloys that had been developed for turbine blades. Peroxide was

The most successful of all the rocket aircraft built by NACA (later Nasa) was the Mach 7 X-15. Seen here is the aircraft in its later X-15A-2 version, with drop tanks for extra fuel, greater endurance and higher speed

used for the reaction control system, and to drive the turbopumps which fed 10,000lb of fuel per minute to the 57,000lb-thrust XLR-99 engine. (Rocket development, therefore, had advanced to the point where the XLR-99 developed the same power as the A-4 engine and used only 60 per cent as much fuel.) NACA acquired a B-52 to launch the heavier new aircraft.

Before the first X-15 was completed in October 1958 it became clear that the Reaction Motors XLR-99 would be late in development. It was decided to speed development by installing a pair of the well tried XLR-11s for "low-speed" tests; the use of quotation marks is advisable, because in May 1960 Joe Walker took the X-15 to Mach 3·19, or 2,111mph, marginally faster than Milburn Apt's fatal speed run in the X-2 four years earlier. With the interim powerplant, the X-15 became the highest-flying aircraft in the world as well, beating Iven Kincheloe's X-2 record with a flight to 136,500ft.

With the XLR-99, the X-15 was expected to operate well beyond the atmosphere. By late 1957 North American had proposed a further step, mating a modified X-15B with the 415,000lb-thrust booster originally designed to launch the Navaho missile. But the project was virtually strangled at birth: following the launch of the Sputnik 1 artificial satellite by the Soviet Union, US work was redirected towards the apparently quicker and simpler technique of reaching orbit with ever larger multistage throwaway boosters, and although the budget for space research rapidly increased there was precious little money available for aerodynamic vehicles of any sort. NACA was renamed the National Aeronautics and Space Administration (NASA) and was given the task of first closing the Soviet lead in space – by putting an American into orbit as soon as possible – and then overtaking the Russians by landing the first man on the Moon.

In the national shock which followed the first flight into space by Yuri Gagarin in April 1961, the progress with the X-15 passed almost unnoticed. The XLR-99 had flown successfully in November 1960, and within a year Bob White had attained the astonishing speed of Mach 6·04, or 4,093mph at 101,600ft. At this speed and altitude the X-15 was encountering stagnation temperatures of more than 1,300°F, rather more than its design limits, and this was to be the highest speed attained by the X-15 for nearly five years. The test programme continued in the quest for greater height, and by August 1963 Joe Walker had flown the X-15 to 354,200ft.

Neither was the X-15 the only potential aerospace programme under way in the US: work was well advanced on a vehicle that nearly turned the Sanger-Bredt project into reality. This project had stemmed from the boost–glide proposals attracted by the MX-2145 bomber study; this had led in January 1958 to a USAF request for proposals for a hypersonic glider, launched by a rocket booster.

Boeing's X-20 Dyna-Soar came close to flight testing in the early 1960s. Note the bluff, rounded leading-edge and nose profiles, designed to spread stagnation pressures and temperatures over as wide an area as possible. Dyna-Soar was a pure delta with upturned wingtip fins (Smithsonian Institution)

134

Although this was originally intended to be a suborbital system, it was soon realised that the problems of putting the vehicle into orbit were not all that much greater than those involved in a slightly lower objective.

The Air Force saw the vehicle as an operational weapon as much as a research aircraft, with applications in strike, reconnaissance and anti-satellite missions. Boeing and a Bell/Martin team responded to the requirement. The most serious problem – apart from the design of an adequately sized and reliable booster, which was a separate USAF project – was heating. Bell/Martin proposed to use coolant tubes to carry heat away from the nose and leading edges. Boeing, however, concentrated on "refractory" materials: steels and other alloys with a high molybdenum content similar to the materials used in safes to resist thermal cutters. These materials radiate heat almost as fast as they can gain it. To get rid of the remaining heat, Boeing proposed to carry a tank of liquid hydrogen as a combined heat sink and source of fuel for the APU.

In early 1960 Boeing was chosen to develop the vehicle. The term "dynamic soaring" had been coined for its suborbital mode of flight, and so the new space-glider became the X-20A Dyna-Soar. The Dyna-Soar was a small single-seat, delta-winged craft – the delta offered good aerodynamics and lift throughout the performance spectrum – with fins on the wingtips and a blunt-based body. As originally defined, the programme was in three stages: Step 1 was suborbital testing, Step 2 would be orbital flights with a larger booster, while Step 3 was the development of an operational military aerospace vehicle. It was planned to achieve the first suborbital flight in 1963, but in late 1962 the US Air Force decided to go straight to the orbital configuration, using a booster derived from the Martin Titan ICBM but featuring large solid-fuel rocket boosters attached to the sides of the liquid-fuelled stage. (This was the first US experience with "parallel staging" in booster design, and the booster is still used as the Titan III.)

In 1963 the United Nations reached agreement limiting the military uses of space; this was the end for the Dyna-Soar, since it eliminated the military requirement for the craft and NASA had other objectives. The programme was finally cancelled at the end of the year, when it was planned to fly the first mission in 1965.

Meanwhile the X-15 programme continued to gather immensely valuable data on kinetic heating at very high airspeeds. In November 1963 NASA decided to rebuild the second X-15, damaged in a landing accident, to make even higher speeds possible. The agency's plan was to explore the applications of "ablative" thermal coatings to an aerodynamic vehicle; these were coatings that simply melted away at high temperatures, and in doing so absorbed and dissipated heat from the stagnation layer. NASA also wanted to test a new class of

powerplant – the supersonic-combustion ramjet or "scramjet". As its name suggests, the scramjet is an engine in which the airflow never drops below supersonic speed: because of this, it can operate at hypersonic speed without attaining intolerable pressures and temperatures.

The rebuilt aircraft, designated X-15A-2, flew in 1964. It had a slightly lengthened fuselage, containing a liquid hydrogen tank to fuel the experimental scramjets which were to be attached to the ventral fin, and two very large external fuel tanks which were used for initial acceleration and climb and jettisoned at Mach 2. In this form the X-15A-2 achieved Mach 6·33 (4,250mph) at 98,900ft in 1966. In the following year the X-15A-2 was covered with a white ablative finish and fitted with a dummy scramjet under the tail. On October 3, 1967, Maj William Knight took the aircraft to Mach 6·72, or 4,520mph. In the process the X-15 encountered stagnation temperatures as high as 3,000°F, burning the dummy scramjet clean off the fin and damaging the ablative coating to the point where it could not be renewed. The aircraft was retired. It had been hoped to rebuild the third X-15 with a slender delta wing for further scramjet research, but in November 1967 it was lost in a fatal crash.

At the time of the last X-15 flights NASA was about a year and a half away from the first manned landing on the Moon. As the achievement of this politically inspired goal drew near NASA was considering what programme should follow the Apollo project so as to keep some of the industry's expertise alive. One answer, which appealed to the military as well as to politicians, was to develop space stations for multiple uses ranging from reconnaissance to earth resources work. There was one drawback: such an operation would require a great deal of traffic between the earth and the orbiting stations, and would consume a great many disposable launch vehicles. Throughout 1968 NASA studied re-usable launch vehicles in various technically feasible forms. The requirement was more demanding than that for Dyna-Soar, which needed a separate disposable booster to get into orbit, and that for the non-orbital antipodal bomber. However, the vehicles shared the same basic principle of recovery by aerodynamic braking and a conventional landing. There were a number of possible variations on the re-usable launcher. It could be built as a single unit, carrying fuel in external tanks which would be jettisoned as fuel was used. Alternatively the vehicle could be built as two recoverable units, gaining the benefits of multi-stage operation while still recovering the complete system.

By mid-1969 NASA's concept of the Space Shuttle, as the vehicle was now known, centred on a two-stage type with a winged launch and re-entry vehicle, or Orbiter, mounted on the back of a huge winged booster, the combination weighing some 3·2 million lb for take-off. At that point the craft was still sized around the station-support mission, with a maximum payload of some 40,000lb. In the course of 1971, however, the Department of Defense, aiming at objectives other than space-station support, began to request improvements in performance: greater payload, with a maximum of 65,000lb, and better "cross-range performance". The latter parameter is the distance which the Shuttle can fly back to its base after descending from a polar orbit; an aerospace vehicle with poor cross-range performance would not be able to descend from such an orbit except when its path passed close to its base.

These requirements tended to increase the weight and cost of the Shuttle, and well before the programme was given a government go-ahead NASA decided to abandon the manned booster. The final configuration, settled when the contract for both engines and Orbiter was awarded to Rockwell – formerly North American – in July 1972, consisted of an Orbiter, mounted upon a very large external fuel tank which was flanked by a pair of huge solid rocket boosters. The Orbiter would carry the "main engines", burning LOX and liquid hydrogen: NASA had gained experience with this powerful fuel mix in the Moon programme.

Development of a rocket motor ten times as powerful as any previous re-usable engine was one major challenge. Perhaps the biggest single problem, however, was the design of the Orbiter's airframe. Aerodynamic efficiency was vital if the aircraft was to land at a reasonable speed with a retrieved load from space, and still meet cross-range requirements. The same requirement for payload meant that the structure weight was of far more importance than it had been on research aircraft such as the Dyna-Soar. On the other hand, the Shuttle Orbiter would have to withstand enormous temperatures.

Both NASA and the USAF had explored one possible approach to the problem: the development of radical aerodynamic configurations. From the point of view of the "aerothermalist", the best shapes are blunt, with wide, smooth frontal curves which spread the stagnation zone. Starting in the late 1950s, NASA and the USAF designed and tested some of the most peculiar-looking aircraft ever flown: strange wingless blobs called "lifting bodies". The first of these shapes to fly was the M2, devised at NASA's Ames research centre; a low-speed manned version made its first flight in August 1963. But a high-speed rocket-powered version suffered a serious accident in gliding trials in May 1967, and had to be grounded for lengthy repairs and modification. Langley Research Centre's HL-10 configuration was the first lifting body to exceed Mach 1, and in February 1970 attained Mach 1·86. In late 1970 the Martin X-24A lifting body also exceeded Mach 1.

But at the time that the Shuttle Orbiter design was being frozen, the lifting bodies had only just started to fly properly; even the little air-launched experimental aircraft, in the hands of some of the world's most experienced test pilots, were tricky to fly. Partly for this reason, the Orbiter was designed with a conventional wing and fuselage. An unswept wing was considered at an early stage, but was rejected in favour of a delta type; the latter is easier to design with a relatively blunt leading edge and its aerodynamic characteristics are better in establishing a high-deceleration, high-angle-of-attack hypersonic glide. The delta's long leading edge also helps to spread the peak kinetic energy release over a wider area. The final double-delta shape was found to offer better trim characteristics at hypersonic speed than a plain delta, together with the same favourable low-speed vortex phenomena as were exploited in the design of the Concorde. The shape of the Shuttle is much more efficient than that of the X-15 despite its fat, slab-sided, payload-oriented fuselage.

The most remarkable thing about the structure of the Orbiter is that it is so conventional. When airframe temperatures reach between 2,000° and 3,000°F, there is no longer a great advantage in building the aircraft out of titanium, stainless steel or Inconel. Temperatures are already far beyond the tolerance of any metals, even chrome-nickel alloys, and some form of external cooling or insulation has to be used. "High-temperature" metals offer few advantages over light alloy at that point, so the Shuttle Orbiter is built basically of conventional aluminium alloys. The airframe is protected by a variety of materials which cover almost the entire surface. The nose, leading edges and rudder, the hottest parts of the airframe, are made of reinforced carbon-carbon and contain no metal fittings at all. The rest of the airframe is covered with tiles of quartz and other insulating and refracting materials, which keep temperatures down to the 350° tolerance of aluminium alloys.

The Orbiter is controlled by a fly-by-wire system based on a central digital computer; the tremendous increase in the power and speed of the digital computer, combined with the almost unbelievable reductions in weight, power and cooling requirements and cost which have followed the revolution in micro-electronics, have led to the virtual elimination of analogue techniques from modern aircraft. (At the time of writing, for example, a programme is under way to "digitise" the flight control, inlet control and air data systems of the USAF's SR-71 fleet.) A measure of the speed with which electronics have advanced is that the black-and-white TV-type displays which replace conventional instruments in the Shuttle cockpit now appear dated beside the full-colour instruments fitted to the latest subsonic airliners.

In the Moon programme NASA had been working with a budget that was virtually unlimited and a fixed schedule; on the Shuttle the mood had changed, and the budget was expected to be controlled. The technique used for the Moon project, in which NASA developed back-ups for most critical

138

systems, could not be used, so when problems occurred with the tiles, the main engines and the huge insulated fuel tank these led to slippage in the schedule. Although the basic airframe of the Orbiter was completed in 1976, and glide-tested from the back of a Boeing 747 in 1977 (marking NASA's final adoption of the piggyback launch method, pioneered in Europe in the 1930s), it was another four years before the first orbital flight. On April 12, 1981, the Orbiter *Columbia* became the first object to return from space under aerodynamic control. The site for the touchdown could not have been more appropriate: Rogers Dry Lake, one of the long lake-bed runways at Edwards AFB.

In a sense the Shuttle is the last word in high-speed flight; no air vehicle can go faster without becoming a spacecraft. On the other hand the Shuttle does not tackle the same problems as an aircraft. Its existence does not mean that the design of a hypersonic vehicle capable of sustained Mach 5 flight will be easy. It is perhaps possible, but is it likely?

As far as is known for certain, there has been only one attempt to design a hypersonic manned air vehicle since the end of the X-15 programme. This stemmed from the latter days of the NASA/USAF lifting-body programme. In about 1970 the US Air Force Flight Dynamics Laboratory arrived at an apparently promising shape for a high-speed lifting body, which was termed FDL-8. One of the Martin X-24 vehicles was radically modified to represent the FDL-8 – in fact the new shape was built up around the structure of the old aircraft – under the designation X-24B. Longer and more slender than earlier lifting bodies, and resembling a highly swept double-delta wing in planform (although, of course, it had no separate wing or fuselage), the X-24B made its first gliding flights in 1973 and immediately demonstrated far better handling than any previous lifting bodies. It achieved a maximum speed of Mach 1·76, could be flown in turbulence with its stability-augmentation system switched off, and was the only lifting body to be landed successfully on a normal runway. (Unfortunately, perhaps, this was much too late to affect the design of the Shuttle Orbiter.)

These results led the USAF and NASA to propose development of a hypersonic research vehicle, the X-24C, based on the same configuration and aimed at a 5,000mph (Mach 7·4) maximum speed. It would have weighed about 55,000lb for launch, about 40,000lb of which would have been propellant, and would have been powered either by the X-15's XLR-99 or by the more powerful LR-105 of the Atlas missile. At that time (1975) General Electric, Marquardt and AiResearch all had scramjet design studies, and the X-24C was designed expressly to test such engines under its flat belly. Among the powerplant configurations foreseen for the X-24C were "external" scramjets, consisting of open vertical channels beneath the fuselage in

Making its first operational missions as these words appear, the US Space Shuttle has been the biggest technological challenge of the 1970s. The resemblance of the Orbiter to the much smaller Dyna-Soar is interesting, though the Orbiter structure is completely different from any previous high-temperature airframe

Lockheed's designers have studied possible successors for the SR-71, including this Mach 5 (3,350mph) design. Such an aircraft could still use hydrocarbon fuel and titanium primary structure with Inconel in the hottest areas, and the powerplants would be turboramjets. Lockheed may have flown a Mach 6 aircraft in secret in the 1970s (Lockheed)

which compression and combustion took place. The first flight of the X-24C would have taken place in 1981, but for reasons that were never stated the programme was not funded to completion. More recently, a configuration closely resembling the FDL-8 has been seen in impressions of a small reusable manned space vehicle under study by the US Air Force.

Since the mid-1970s, however, there have been persistent but unconfirmed reports that Lockheed's Skunk Works has actually flown a manned aircraft capable of sustained flight at Mach 6, or 3,900mph+. A 1977 Lockheed study shows a Mach 6, hydrogen-fuelled airliner powered by turbojets and scramjets. Hydrogen, in particular, appears suited to such aircraft. Its great energy content per

unit of weight allows high cruising altitudes, and its low storage temperature could solve a great many airframe and system cooling problems.

Will hypersonics be the next wave of development? In the early 1980s we are beginning to sense a thaw in the "speed freeze" that has affected most aircraft since the early 1960s; the YF-104A, flown in 1956, could show a clean pair of heels to most fighters now in production, at any altitude. Attitudes are changing now; technology which stemmed from the US supersonic transport programme is leading to military "supercruise" designs capable of sustained high altitude and high-Mach performance. And when the fighters get up there we are back to the old answer for the reconnaissance and strike aircraft: fly higher, and fly faster.

In the late 1980s the US Air Force may introduce a smaller reusable space vehicle to complement the Shuttle. The basic system would be unpiloted, but a manned version could develop into a "space fighter" for anti-satellite and reconnaissance missions *(Boeing)*

Index

143